The Surveyor's Expert Witness Handbook: Valuation

Martin Farr

2005

Books

A division of Reed Business Information

Estates Gazette
1 Procter Street, London WC1V 6EU

©Martin Farr, 2005

ISBN 0 7282 0463 0

Typeset in Palatino 10/12
Printed by Bell & Bain Ltd, Glasgow

Contents

Acknowledgements

In the preparation of this book, I wish to acknowledge with thanks the valuable assistance given to me by Nicholas Cheffings of Lovells. I also wish to thank Colin Passmore of Simmons & Simmons, Ian Padbury of the RICS Dispute Resolution Service, Brian Thompson of the Expert Witness Institute, Rebecca Copley of Eversheds LLP, Don Baker in the City Valuation Office (Valuation Office Agency) and Jonathan Ross of Forsters.

Introduction

In our 21st century consumer-led society, there is an increasing belief, indeed often the firm expectation, that when things go wrong, a remedy is available, if necessary through the courts. There is also continuing growth in the development of appeal rights associated with statute-based intervention, which must be seen to impact widely, fairly and correctly. The resulting trend towards increased litigation is played out against the backdrop of a society grappling with ever more complex, technical, economic and social issues given wider circulation and exposure by our advanced information-based technology.

It is, therefore, no accident that the need for expert evidence and advice in dispute resolution has steadily increased. The field of property valuation is no exception. Commercial and residential property valuers have experienced increasing demands for their services from a wide range of sectors and for differing purposes.

The range of disputes with property value-related issues include actions in the High Court, county courts, compensation and rating appeals before the Lands Tribunal, leasehold enfranchisement proceedings, commercial property lease renewals, commercial and industrial property rent reviews before arbitrators and independent experts, rating and council tax appeals before local valuation tribunals and so on. This list is not exhaustive.

The object of this book is to provide the commercial property valuer with a general introduction to providing expert evidence in a litigation context, and the rules, requirements and the pitfalls for the unwary. Particular trouble has been taken to emphasise the need for quality

1

evidence based on relevant experience that is objective, unbiased, independent and of sufficient quality to resist challenges before the courts, tribunals and arbitrators.

It is my experience that surveyors often do not fully understand the need for a wholly unbiased approach. The temptation to persuade, exaggerate or veer towards advocacy, fulfilling the natural desire to please those who are paying the fee, can often result in flawed evidence, which, ultimately, may be detrimental to the client's case.

Within the scope of this book, it is impossible to provide a detailed exposition of all procedures, relevant valuation and technical issues as to do so would fill a library. Nevertheless, it is hoped that this handbook will help the practitioner to start off on the right course and be forewarned of the issues that he* is likely to face. As always, there is no substitute for practical experience but it is hoped that, as a result of this handbook, greater professional awareness and interest will be generated, leading to higher standards of valuation expert evidence in all commercial property fields. The valuer will find that experience can be supplemented by the many courses held from time to time on specific areas of expert evidence practice, including legal updates and procedures, report writing and courtroom skills training.

On the rare occasion when a settlement has not been achieved, expert evidence needs to be reported and tested before a court, arbitrator or tribunal. While attendance at a hearing is now rare, it must always be anticipated. It can be challenging but not always the ordeal that is widely feared if the right steps are taken in accordance with the advice in this handbook.

If I have managed to convey the basic elements of sound, well-presented, objective evidence prepared in accordance with relevant procedures, this may have started the aspiring valuation expert down a satisfying road in terms of professional development, while contributing to fairer justice and a satisfied client.

---

* Wherever "he" or "his" is used in this book, this refers to "he or she" or "his or her".

Background

2

The majority of well-established judicial systems throughout the world rely, to a greater or lessor extent, upon expert evidence. The role of the expert in advising on specialist, scientific or technical matters makes a vital and fundamental contribution towards the fair and proper administration of justice. Judges and tribunals will frequently not have sufficient knowledge or experience in such areas in order to reach fair judgments awards or determinations.

It will be advantageous that a tribunal or arbitrator will often have some specialist knowledge or experience of particular technical issues, but rarely will these be sufficient to reach a thorough understanding of the particular circumstances portrayed in evidence. The expert will have the ability and opportunity to concentrate on the issues in depth. In my experience, many High Court judges rarely have more than a very elementary grasp of the art of property valuations and the associated technical issues. Indeed, they should rarely be expected to have such an understanding.

Expert evidence in civil actions

It must be emphasised, however, that in civil actions expert evidence is not admissible as of right. Section 3 of the Civil Evidence Act 1972 may allow for the introduction of such evidence. It has been held in *Barings* v *Coopers & Lybrand (a firm)* [2001] EWCA Civ 1163, a case concerning the much-publicised unauthorised trading by Nick Leeson, that it will be admissible under Section 3: "In any case where the court accepts that there exists a recognised expertise governed by recognised standards and rules of conduct capable of influencing the court's decision on the issues which it has to decide".

It is always important to bear in mind that the expert's role is to provide evidence on the necessary technical issues so that it is fully understood by the court. It can then be evaluated and given an appropriate degree of weight in the formulation of a decision. The expert evidence itself will not provide the judicial conclusion.

Over the past 50 years and into the millennium, three important but connected trends may be perceived in the development of civil dispute resolution in England and Wales and all of which have important implications for the valuer expert witness.

Growth of dispute resolution

There has been significant growth in nearly all areas of dispute resolution. The exercise of democratic rights of appeal grow as the obligations of individuals and businesses develop in reaction to the ever-growing statutory interference and regulation of their affairs. In addition, this trend is given added impetus as our consumer society continues to indulge in what has been called "the blame culture" of seeking compensation for perceived injury, financial loss or whatever, if necessary, via the courts.

By way of illustration, the number of non-domestic civil actions registered in the High Court within England and Wales for 2003 indicate that this had grown to 1.790,000. However, the promotion of alternative dispute resolution has caused the number of court hearings to decline quite dramatically, in the case of the Queens Bench Division of the High Court, from 120,000 in 1998 to 14,000 in 2003 (71,300 for all non-domestic civil actions).

The Valuation Office Agency records indicate that the number of appeals lodged in the 1990 Rating Revaluation was 1.539 million, compared with 1.508 million in the 2000 Rating Revaluation. Interestingly, however, a smaller proportion of appeals (1%) were determined by the local valuation tribunals, compared with those brought to a hearing at the earlier Revaluation.

In the field of rent review dispute resolution, the RICS Dispute Resolution Service received 3828 applications in 1996 for the appointment of a rent review arbitrator or expert, compared with 7942 in 2003 (59% were for arbitrator). It has to be said that the number of appointments has since declined due to the current changed market conditions, particularly in the offices sector. It is to be expected that the number of appointments will eventually revive and increase over time when market conditions improve and more rent reviews are "triggered" and negotiated.

Thus, the need for expert valuation evidence and advice is likely to increase further in line with overall trends but with some ebb and flow within individual cycles, such as the quinquennial Rating Revaluations and market-induced changes. It is particularly noticeable, for example, that some forms of civil litigation tend to increase in depressed market conditions, no doubt reflecting the greater need for a remedy when business losses are most prevalent. However, the nature of the role will change as dispute resolution practice develops.

Dispute resolution by tribunals

The second most obvious trend is towards the much increased use of specialist tribunals and other means of alternative dispute resolution away from the courts, at least at first instance, but with the ability to appeal to the High Court or beyond, at least on a point of law.

The use and occupation of commercial property is subject to taxation, both local and national, and is otherwise widely regulated by statute. Rights of appeal or other remedies have spawned a large number of alternative dispute systems and bodies, some of which are now long established.

Rating appeals

Surveyors will be familiar with the long-established rating system of local government taxation, rates being levied on assessments prepared by the Valuation Office Agency in accordance with defined rental valuation criteria and which are now revised every five years.

If not resolved, appeals are heard by independent local valuation tribunals, now administered by the Valuation Tribunal Service. Where legal issues exist, or the case is of sufficient size or complexity, appeals can be made to the Lands Tribunal, an independent body with High Court status, the members being experienced senior QCs or surveyors.

The Lands Tribunal also has referred to it applications made for the assessment of compensation for compulsory purchase, restrictive covenant disputes within the provisions the Law of Property Act 1925 and a number of other appeal rights arising out of other specific statutory provisions.

Rent review arbitration

With the emergence of significant rates of monetary inflation from the late-1960s, it became the practice to insert rent review clauses in commercial property leases, initially on seven or 14-year cycles, but later reduced to five-year intervals. The rent review clause was (and is) usually based on the concept of a hypothetical open market letting between a willing lessor and a willing lessee. The precise interpretation of such clauses has generated a considerable amount of litigation. If the parties cannot agree the review, it is normally provided that it should be referred to an arbitrator or independent expert appointed, in the absence of agreement between the parties, by the president of the Royal Institution of Chartered Surveyors (RICS).

Since it is normal for valuer experts to be appointed by both landlord and tenant in such a dispute to give evidence to the arbitrator, this represents a considerable area of expert evidence work that has developed over the years in relation to the various commercial and industrial property sectors.

In recent years, it has, in addition, become the practice on occasion to refer the Landlord & Tenant Act 1954 commercial lease renewals to a RICS-appointed arbitrator in preference to a hearing before the local county court. The attractions being the arbitrator's greater specialist expertise, speed and possibly lower costs. These referrals are made under what have become known as the PACT (Professional Arbitration on Court Terms) rules, a joint initiative of the RICS and the Law Society.

Mediation

Within the overall trend, alternative dispute resolution has embraced the concept of mediation. It must be distinguished, however, from other means of dispute resolution. The role of the mediator is to persuade the parties to find common ground in a dispute that may lead to a settlement. He is there to try and narrow the issues between the parties by assuming the characteristics of a "go between" or exercising what has often been termed, on the world stage, "shuttle diplomacy". It is not about a judgment of the evidence.

Under the Civil Procedure Rules discussed later, judges commonly direct that mediation is to be attempted as part of the ongoing procedures in order to try and avoid the costs of a trial. From the most recent data available, it appears that around 96% of High Court civil actions commenced are resolved without resorting to a trial hearing and more than 50% of these have been the result of a reference to mediation.

In addition to the ever-increasing volume of litigation cases and the trend towards various forms of alternative dispute resolution, important developments have emerged in the methods of securing justice.

Adversarial versus inquisitorial

Our legal system relies on an adversarial basis of presenting and testing evidence, as opposed to the inquisitorial system favoured in continental Europe. Until recently, the judge's role was mainly to listen to the conflicting evidence put by the parties on their behalf, both expert and of fact and to determine which case more closely resembled the truth. He would only intervene to ensure that the rules of evidence were properly applied, to seek clarification and to ensure "fair play".

Many have questioned whether justice is best served by deciding a case on the basis of a contest in which the better-presented evidence and the greater skills of cross-examination may be overwhelming factors in persuading the judge or judicial body. In contrast, criticisms have been encountered by those who believe that the inquisitorial system relies too heavily on the judgment of one person, since only the examining judge or magistrate conducts the investigations, calls witnesses, and so on. The system could be open to abuse and may call into question standards of integrity and independence.

Despite these criticisms, the adversarial system largely remains but there is a trend towards a more inquisitorial approach by judges. This is mirrored by the greater use of tribunals and arbitrators, who often have greater specialist knowledge and expertise. They are able to adopt a more "hands on" approach, with greater control of the process and a greater involvement in asking questions of witnesses.

An example of this trend is the power given to arbitrators under the 1996 Arbitration Act to undertake their own investigations as to fact, unless the parties have agreed otherwise. In my experience as an arbitrator, however, the ability to "step down into the arena" is rarely needed, the threat of such action is usually sufficient to produce the information, further evidence or clarification.

It must be emphasised, however, that the methods of testing evidence by cross-examination remain and if undertaken fairly within the rules, the judge, arbitrator or tribunal will not (normally) intervene.

Civil Procedure Rules

The emergence of the above trends naturally leads me on to the most important event in the recent history of administering civil justice in the courts. As a result of the volume of cases, the perceived high costs of administering the system, hitherto largely dictated by the parties, and seemingly to a degree the sometimes perverse limitations of the adversarial system, the then Master of the Rolls Lord Woolf was asked to commence an inquiry into the system of civil justice in England and Wales in 1994. He produced an interim report entitled *Access to Justice* in June 1995 and a final report in July 1996. He concluded that civil proceedings were often too slow, complex, costly and, to a degree, open to some abuse (not least in the realm of expert evidence) and sought to remedy these problems.

The resultant Civil Procedure Rules, usually referred to as CPR, were published in January 1999 and a considerable number of amendments have since been made. Updated rules as a whole can be found on the Lord Chancellor's website at *www.lcd.gov.uk*.

The rules in part 1 give the "overriding objective" of CPR. As far as the expert witness is concerned, they are of sufficient and fundamental importance to be repeated in summary from as follows:

CPR Part 1 — the overriding objective

"... is to enable the Court to deal with cases justly."

The following issues are relevant to how the Court will ensure that cases are dealt with "justly", a task which must be borne in mind at all times when the Court is giving effect to the new procedures and also exercising its discretion:

* Ensuring that the parties are on an equal footing.
* Saving expense.
* Dealing with the case in ways which are proportionate:
    (i) to the amount involved
    (ii) to the important of the case

    (iii) to the complexity of the issues

    (iv) to the financial position of each party.

- Ensuring that it is dealt with expeditiously and fairly.
- Allotting to it an appropriate share of the court's resources.

The parties have a duty to assist the court in reaching these objectives. The court has general case management powers, which far exceeded those previously applied by judges.

It is important to emphasise, in particular, the doctrine of "proportionality". It is no longer possible, to take an example, for a party to indulge in an expensive QC and expert witnesses in a relatively minor case. The parties must be perceived to be on an equal footing. The case must proceed in a manner proportionate to the amount of money and circumstances.

There has to be a far greater degree of disclosure and transparency. The parties must co-operate in regard to narrowing issues and deciding which need further investigation. Thus, for example, the previously common gamesmanship in holding back important evidence, with the aim of planning a defence "ambush" in court, is no longer to be permitted. The parties must "put their cards on the table".

Pre-trial procedures will be supervised by the judge using case management powers and the court will, therefore, have greater control over the preparations for presenting evidence.

Opinions are mixed as to the precise impact of CPR. It would seem that there has been relatively little impact upon costs but, to a degree, delays in the court system have been eased. Undoubtedly, the rules have established a far better pre-trial environment for parties to draw closer in the consideration of their disputes at an earlier stage. Far fewer cases reach trial since there has been a greater willingness and openness in the pursuit of possible settlements.

There has been a substantial impact upon the way in which expert witnesses are expected to conduct themselves in terms of independence and impartiality. Generally, they display less of the "hired gun" mentality characteristics much criticised by Lord Woolf. However, in my view, there is still often considerable room for improvement.

I will return to more detailed aspects of CPR and the procedures as they affect the valuer expert witness in chapter 4.

General Principles

The role and duties of the expert witness have been well defined in the well-known judgment of Cresswell J in *National Justice Compania Naviera SA* v *Prudential Assurance Co Ltd* [1993] 2 EGLR 183 (simply referred to in practice as "the Ikarian Reefer"). Repeated verbatim (including the judge's reference to precedents), the duties and responsibilities of expert witnesses in civil cases include the following:

- Expert evidence presented to the court should be, and should be seen to be, the independent product of the expert uninfluenced as to formal content by the exigencies of litigation (*Whitehouse* v *Jordan* [1981] 1WLR 246 on p256 Lord Wilberforce).
- The expert witness should provide independent assistance to the court by way of objective, unbiased opinion in relation to matters within his expertise (see *Polivitte Ltd* v *Commercial Union Assurance Co plc* [1987] 1 Lloyds Rep 379 on p386 *per* Mr Justice Garland and *Re J* (1990) FCR 193 *per* Mr Justice Cazalet). An expert witness in the High Court should never assume the role of an advocate.
- An expert witness should state the facts or assumptions upon which his opinion is based. He should not omit to consider material facts that could detract from his concluded opinion (*Re J* sup).
- If an expert's opinion is not property researched because he considers that insufficient data is available, then this must be stated, with an indication that his opinion is no more than a provisional one (*Re J* sup). In cases where an expert witness who has prepared a report could not assert that the report contained the truth, the whole truth and nothing but the truth, without some

qualification, the qualification should be stated within the report (*Derby & Co Ltd* v *Weldon* [1991] 1 WLR 652 *per* Lord Justice Staughton).

- If, after exchange of reports, an expert witness changes his view on a material matter having read the other side's expert's report or for another reason, such change of use should be communicated (through legal representatives) to the other side without delay and, when appropriate, to the court.
- Where expert evidence refers to photographs, plans calculations, analysis measurements, survey reports or other similar documents, these must be provided to the opposite party at the same time as the exchange of reports.

The Civil Procedure Rules in Part 35 further reinforce these principles. Part 35.3 states that:

- It is the duty of an expert to help the court on the matters within his expertise.
- This duty overrides any obligation to the person for whom he has received instructions or by whom he is paid.

Anglo Group plc v *Winther Brown & Co Ltd* (2000)

The judgment of Toumlin J in *Anglo Group plc* v *Winther Brown & Co Ltd* (2000) Info TLR 61 further expands and updates the general duties of an expert witness taking into account the Civil Procedure Rules. The further refined principles are worth quoting in full (the headings are my own) as follows:

- **Objective unbiased opinion — not advocacy**
  An expert witness should, at all stages in the procedure, on the basis of the evidence as he understands it, provide independent assistance to the court and the parties by way of objective, unbiased opinion in relation to matters within his expertise. This applies as much to the initial meetings of experts as to evidence at trial. An expert witness should never assume the role of an advocate.

- **Technical and professional testimony — not judgmental**
  The expert's evidence should normally be confined to technical

matters on which the court will be assisted by receiving an explanation, or to evidence of common professional practice. The expert witness should not give evidence or opinions as to what the expert himself would have done in similar circumstances or otherwise seek to usurp the role of the judge.

- **Narrow the issues**
  He should co-operate with the expert of the other party or parties in attempting to narrow the technical issues in dispute at the earliest possible stage of the procedure and to eliminate or place in context any peripheral issues. He should co-operate with the other expert(s) in attending, without prejudice, meetings as necessary and in seeking to find areas of agreement to define precisely areas of disagreement to be set out in the joint statement of experts ordered by the court.

- **Independent evidence**
  The expert evidence presented to the court should be, and be seen to be, the independent product of the expert uninfluenced as to form or content by the exigencies of the litigation.

- **Assumption — scope of differing opinions**
  The expert witness should state the facts or assumptions upon which his opinion is based. He should not omit to consider facts that could detract from his concluded opinion.

- **Outside the expertise**
  An expert witness should make it clear when a particular question or issue falls outside his expertise.

- **Inadequate facts**
  Where an expert is of the opinion that his conclusions are based on inadequate factual information, he should say so explicitly.

- **Impact of new facts or opinion**
  An expert should be ready to reconsider his opinion, and, if appropriate, to change his mind when he has received new information or has considered the opinion of the other expert. He should do so at the earliest opportunity.

Independent, unbiased opinion

It may be imagined that the above principles appear succinct, clear and unambiguous, leaving little room from differing practice. Unfortunately, in my experience, some surveyor expert witnesses have not fully grasped the rationale and philosophy of independent, unbiased opinion and duty to the court or other "judicial body" in all matters.

There is a natural tendency, leading to inbuilt conflict, for any professional to wish to please his clients, by which is meant those who are paying his fees. Perhaps because most surveyors practice in a commercial environment giving advice and negotiating transactions, there is thus a natural tendency to wish to continue to act in the supposed best interests of the client. This leads to the mistaken belief that a partisan approach can be adopted in the realm of providing expert evidence.

At its extreme, this is not only in breach of court rules but may, in rare circumstances, be actionable or at least become a matter that could be reported to the professional institution of the valuer, resulting in disciplinary proceedings. In addition, as Mr Justice Toumlin noted in his judgment, failure to take an independent approach may ultimately have the effect of inflating the costs of the parties. Reference will be made to *Phillips* v *Symes* [2004] EWHC 2330 (in chapter 14 on Expert witness immunity) where it was held that an expert witness who is in breach of the duty to the court can be liable for the resulting costs unnecessarily incurred or "wasted".

Attempts to colour evidence to the supposed advantage of the client may also, in my experience, backfire and have the opposite result to the one intended. I have been the arbitrator in numerous cases where it has become quite clear that an expert's testimony is coloured, selective, exaggerated or even just plain dishonest in the pursuit of providing the best "spin" for the benefit of the party on whose behalf the evidence is presented. This often results in either complete or significant loss of credibility. The degree of weight that is attached to such evidence may suffer in consequence and the client's case may suffer as a result.

Surveyors and valuers should remember that when giving evidence orally before a tribunal or arbitrator or in the High Court, attempts, in effect, to pervert the course of justice by exaggerated, selective or dishonest evidence may have the most disastrous effects under skilled cross-examination. Serious doubts over credibility may well emerge.

Experts who appear not to be consistent in their advice or evidence can expect to be subject to attempts to discredit them. In *London & Leeds Ltd* v *Paribas* (No 2) [1995] 1 EGLR 102, an appeal from a rent review arbitration succeeded in forcing the disclosure of evidence given in another rent review arbitration by the same surveyor expert as being admissible, if he had expressed himself differently in the earlier case.

In *Royal & Sun Alliance Trust Co Ltd* v *Healey & Baker* 2000 WL 1741454, Mr Justice Hart had some harsh comments to make on the valuation expert evidence presented, in particular on behalf of the claimant. He agreed with the defence that the witness was "extremely selective" in the adaptation of evidential material in support of his opinions, including reference to IPD indices, economic property reviews and industrial letting evidence. The selectivity became apparent under cross-examination and, as a result, the case for falling rental values was not made.

Questions of expertise and independence were aired in *SPE International* v *Professional Preparation Contractors Ltd* [2002] EWHC 881. The judge criticised the expert for his lack of expertise and found that he was not independent since he had previously acted as a management consultant to the claimant. His evidence was inadmissible.

Expert opinion and advocacy

There is also a temptation to argue points in an expert's report, rather than merely provide opinion, which is the only duty of the expert witness. In my opinion, it can be extremely damaging to the credibility of an expert, who may, quite naturally, be viewed by the court, arbitrator or tribunal as merely acting as an advocate. As a result, any elements of expert opinion are debased or may not be viewed at all as expert testimony.

It cannot be advised too strongly that the expert witness valuer should resist all temptations to provide other than his independent, unbiased, objective, honest opinion or opinions, presented in the form which will give the most assistance to the court or tribunal without resorting to arguing points.

In *Gareth Pearce* v *Ove Arup Partnership* 2001 WL 1251820, an architect for the claimant's expert witness indulged in advocacy. Mr Justice Jacob, at first instance, found that the case as claimed was on the evidence of the expert architect "pure fantasy". The judge was very critical of the biased and irrational evidence given. The expert had come to agree the case and his evidence was not substantiated. He bore a heavy responsibility for the case coming to trial. It was held that the evidence be discounted and the only sanction open to the court was to refer the architect's conduct to his professional body. You have been warned!

If there is a need, for example perhaps in written submissions to an arbitrator, to both perform the advocacy role and provide expert opinion, the two functions should be kept quite distinct and separate. The arbitrator should be left in no doubt as to when the submission is pleading the case with arguments in favour or criticisms in rebuttal and when the independent, unbiased opinion of the expert is being given.

It has often been put to me by even experienced surveyors that they find it difficult to convert from the role of a professional adviser and negotiator to one of providing independent expert witness evidence. It is accepted that the culture conversion may be difficult but it must be faced and overcome not least in order to inform those instructing the surveyor.

Clients will often wrongly expect that the previous role of acting in their best interests in a negotiating sense will continue in pursuit of persuading the court or tribunal to reach the desired determination. Therefore, it is important either directly, or through the retained solicitors if appointed, to explain to the client the changed philosophy that is now required. This can be often achieved at the appropriate time in formulating conditions of engagement. This issue will be referred to in later chapters. The earlier it is done the better. A client who perceives the expert to be a "hired gun" needs to be put right at the outset.

Experience and expertise

In my experience, the second most common fault of valuation experts is to provide expert evidence beyond their established area of expertise. Qualifications in themselves are not enough. It is obvious, but worth restating, that the expert witness valuer must ideally have had at least some years experience of the particular type of property and property market in respect of which he is required to give his opinions. It is accepted that, on occasion, this may be difficult to define, particularly in the case of more specialist properties, which are infrequently encountered.

Perhaps the acid test should be whether a well-informed substantial commercial client would instruct you otherwise to give an opinion on such a property, or would your peers in the same professional field regard you as an acknowledged expert to turn to for a second opinion?

It is not sufficient to have some knowledge of the property type, but to have been deeply involved with such properties and perhaps also within the locality concerned within a reasonable period of instructions being received.

It has concerned me greatly in recent years to encounter "experts" at expert witness conferences in a number of fields, who regularly assume the role of an expert witness notwithstanding that they have not had actual "hands on" professional experience for some years. This can often be the case, for example, in the medical world; where the "professional expert" can be encountered!

In the property field, this means the expert must have had recent and continuing experience, whether as a deal maker or a valuer, in the particular market within which the subject property falls. This also means being well versed in a general sense in regard to relevant background issues, for example, planning, landlord and tenant, and environmental compliance.

Some years ago, I was appointed as arbitrator on a higher education rent review in the London area. It became clear from the evidence presented to me on behalf of the lessee that the expert concerned had had little or no experience in the particular sector and was wholly ignorant of the then current educational trends and related market value issues. Very little weight was attached to his evidence.

It is self-evident that the expert has a duty to explain matters fully to the court or tribunal since this will facilitate a clearer understanding of the evidence, thus indirectly leading to enhanced credibility. However, in my experience, this is not often fully understood. The issues concerned will be further considered in later chapters.

Agreeing facts

There is an important overall responsibility to agree as many matters of fact as possible and if there are differences of opinion in the experts' reports to narrow these as far as possible. In the early stages of a legal action, those instructing expert witnesses (whose task is to secure maximum tactical advantage) are often, understandably, nervous that any prior contacts between experts in advance of evidence being exchanged might compromise a client's position. Nevertheless, the expert's duty is to the court. If time and money are to be saved, it is up to the expert to persuade those instructed that the overriding duty to the court takes precedence. These issues will be further discussed later.

Conflicts

The important question of conflicts of interest will be considered later but it will suffice to say, at this juncture, that the expert witness must be wholly independent and not have had any relationships that would give reasonable grounds for believing that he may be open to bias. In this context, the difference between the "admissibility" of evidence and the weight to be attached to it is important, as is explained later.

Non-partisan evidence

A common fault of expert witnesses is not to appreciate that the fundamental duties and obligations mean going beyond the evidence being put and how it is formulated. If there is room for differing opinion or if there are facts that could, in the opinion of another, possibly lead to a different conclusion, the expert must assist the court of tribunal by exploring these and give reasons why the evidence is to be preferred.

Credibility will be enhanced by an even-handed approach, providing a broad, balanced view having considered all the alternatives and being totally open in regard to possible alternative opinions. Obvious integrity earned by such a broad authoritative approach will tend to score high with the judge or arbitrator. However, it will be obvious that if this results in any serious dilution of the opinion given, then perhaps the alternative basis of expert evidence should have been the preferred conclusion in the first place! Concessions may enhance credibility but not if they are too frequent or too fundamental.

A good example of a willingness to change an opinion following discussions between experts occurred in *Transco plc* v *Griggs* [2003] EWCA Civ 564. Both experts revised their opinions to some degree after two meetings. The Court of Appeal said that an expert's willingness to change his opinion as discussions take place and the evidence develops may be a sign of strength not a weakness.

Court Rules and Codes of Practice

Civil Procedure Rules

The Woolf report and the resultant Civil Procedure Rules (CPR) are now considered in a little more detail as they affect the role of the expert witness.

It is worth a reminder at this stage of the principal changes brought about by the Woolf reforms as follows:

- More transparency between parties and a trend away from the purely adversarial culture.
- New High Court and county court rules.
- New procedures and forms of pleadings.
- Clearer annunciation of issues and less jargon.
- Greater powers for the courts to control the management of cases.
- Allocating cases to three tracks.
- The ability to make cost orders against parties who are in breach of the new rules.
- Stricter timetables.
- Limitation on disclosure.
- Greater control on the use of expert evidence, with stricter guidance for the expert's independent role and duty to the court and a greater use of single joint-appointed experts.

Case management

It is provided in CPR Part 1.3 that the parties are required to help the court to further the overriding objective (see page 12 of CPR). The court has a duty to manage cases actively and in CPR Part 1.4 active case management is said to include:

- Encouraging the parties to co-operate with each other in the conduct of the proceedings.
- Identifying the issues at an early stage.
- Deciding promptly which issues need full investigation before trial and accordingly disposing summarily of the others.
- Deciding the order in which issues are to be resolved.
- Encouraging the parties to use an alternative dispute resolution procedure if the court considers that appropriate and facilitating the use of such procedure.
- Helping the parties to settle the whole or part of the case.
- Fixing timetables or otherwise controlling the progress of the case.
- Considering whether the likely benefits of taking a particular step justify the cost of taking it.
- Dealing with as many aspects of the case as it can on the same occasion.
- Dealing with the case without the parties needing to attend at court.
- Making use of technology.
- Giving directions to ensure that the trial of a case proceeds quickly and efficiently.

Many of the above strictures will directly or indirectly affect the conduct of the expert witness, particularly in regard to identifying the issues at an early stage, the co-operation between the parties, the use of alternative dispute resolution (mainly mediation) and so on.

All the rules themselves are supplemented, in some instances, by practice directions and reference will be made to these where relevant to the expert witness' role. There is also an important Code of Guidance protocol, which is discussed later.

The case management "tracks"

CPR Part 26, which deals with case management, provides for three case management tracks as follows:

- The small claims track, which is generally in respect of claims having a financial value of less than £5000 and certain other claims, for example, for repair work between the landlord and tenant having a limit of approximately £1000.

- The fast track is for claims having a financial value of not more than £15,000 but only where the court considers that the trial will not last more than one day and oral expert evidence will be limited to one expert per expert field, there being no more than two expert fields.
- The multi-track will be generally for any other claims not covered in the foregoing. In the majority of commercial property value cases reaching the courts, it is to be expected that these will be multi-track, whether in the county courts or the High Court. Claims for less than £50,000 will generally be tried in a county court under the appropriate practice direction.

CPR Part 35 — experts and assessors

Part 35 of the CPR concerns experts and assessors. Given its importance for expert witness valuers, the whole of CPR Part 35 is reproduced in appendix I. Some general principles can first be drawn from CPR Part 35 and these are:

- Expert evidence should be restricted to that which is reasonably required to resolve the proceedings (Part 35.1).
- It is the duty of the expert to help the court on the matters within his expertise. This duty overrides any obligation to the person for whom he has received instruction or by whom he is paid (Part 35.3 (1) and (2)).
- No expert evidence can be called without the court's permission (Part 35.4 (1)).
- The court can limit the amount of the expert's fees and expenses, which are recoverable from any other party (Part 35.4 (4)).
- Expert evidence must be given in a written report unless a contrary direction is given. However, in the case of a fast-track claim, the court will not require an expert to attend unless it is necessary to do so in the interests of justice (Part 35.5 (1) and (2)).

Cost limitations

It is important to bear in mind that the costs of the expert, which are often on a time basis, may have been agreed with the instructing solicitor on behalf of his client. This may be called into question at a later stage in the proceedings if the issue of costs issues becomes paramount, particularly if there is an application to the court for limitation, or the costs become subject to taxation. The client will not thank the expert for excessive time charging of fees if, ultimately, he is unable to obtain recovery from the other party (as a result of the judgment of the court), or where costs have in any event been awarded against.

The expert's report and privilege

There are stipulations in regard to the contents of an expert's report and these include:

* A statement that the expert understands his duty to the court and has complied with that duty (Part 35.10 (a) and (b)).
* The report must state the substance of all material instructions, whether written or oral, on the basis of which the report was written (Part 35.10 (3)).
* Instructions to give a report are not privileged (that is, against disclosure) but the court will not order disclosure of any specific document or permit any questioning in court other than by the party who instructed the expert unless there are reasonable grounds to consider the statement of instructions to be inaccurate or incomplete. (Part 35.10 (4) (a) and (b)).
* The report must comply with the requirements of the relevant practice direction (see below).

The practical requirement in the third point above is important. Interestingly, it is unnecessary to include within the report the actual letter of instruction, which may refer to other possibly privileged matters, but only "the substance" of the instruction given to the expert. However, it should be borne in mind that if it transpires there is a "hidden agenda" in terms of the instructions that were given, there is no privilege from disclosure. In my experience, it may be necessary to discuss this issue in advance with the instructing solicitors if, in the circumstances, this seems to be advisable, particularly where the expert provides advice in the early stages to the intending litigant (or its legal advisers). The question of privilege is covered in more detail in chapter 15.

In the practice direction, which supplements Part 35, the required form and content of an expert's report is further amplified. It is vital that the issues referred to are understood and that the requirements are complied with. The substance of the practice direction is therefore repeated. It requires that an expert's report must:

- Give details of the expert's qualifications.
- Give details of literature or other material relied upon in the report, that is to say who carried out any test or experiment.
- Where there is a range of opinion on the matters dealt within the report, summarise the range of opinion and give reasons for his own opinion.
- Summarise the conclusions reached and provide a statement that the expert understands his duty to the court and has complied with that duty.
- Obtain a statement setting out the substance of all material instructions (whether written or oral). This must include the facts of the instructions given to the expert that are material to the opinions expressed in the report or upon which those opinions are based.
- The report must be verified by a statement of truth as follows:
  I confirm that insofar as the facts stated in my report are within my own knowledge I have made clear which they are and I believe them to be true, and that the opinions I have expressed represent my true and complete professional opinion.

I have already referred to the importance of "balance" in the report and the need to address all possible ranges of opinion, giving reasons for the preferred conclusions.

Agreeing facts and narrowing the issues

It is provided in Rule 35.12 that the court may, at any stage, direct that a discussion takes place between experts for the purposes of requiring them to identify the issues in the proceedings and, where possible, reach agreement on any issues. In addition, the court can specify the issues that the experts must discuss and may direct that, following a discussion between the experts, they must prepare a statement for the court showing those issues on which they agree and those where they disagree, with a summary of their reasons. However, it is specifically provided that the discussion between experts should not be referred to at trial unless the parties agree.

It is essential to the overriding objective of CPR that experts do try and narrow the diverging issues between them and agree facts as far as possible. In my experience, this can cause difficulty in practice, sometimes more than any other in the overall role of the expert witness, particularly when instructing solicitors are involved.

As an arbitrator it is, in my experience, standard practice to direct that the facts and, as far as possible, comparable transactions upon which the expert valuers intend to rely are to be agreed in advance. Unfortunately, the importance of prior agreement in cases of property valuation is so often lost on those instructing valuation experts in other forms of dispute resolution. Considerably increased costs can be suffered in practice where the opportunity is not taken at a very early stage of proceedings for the experts on each side to at least agree facts and attempt to narrow the issues. In property valuation disputes, the facts and assumptions will often determine the basis and amount of the value opinion.

It has been my experience that instructing solicitors (in the absence of a suitable court direction) are often reluctant to commence this process for fear of prejudicing or surrendering their independent preparation of the case for trial or negotiating strategy. I will return to this issue later in regard to experts' meetings and the way in which this problem may be overcome. However, under CPR the judge has the ability to ensure the experts meet early rather than late, hopefully before experts' reports are exchanged and not after. The expert has the right to ask for court directions within Rule 35.14 but, in practice, it will be a brave or foolhardy expert who makes use of this right without the consent of his instructing lawyer.

Single joint experts

Rule 35.7 gives the court power to direct that evidence be given by a single joint-appointed expert. Where the parties cannot agree on the identity of the expert, the court may select the expert from a list prepared by the instructing parties or direct that the expert be selected in another manner. I have never experienced a situation where the court has made such an appointment, but I have acted as a single joint-appointed expert where the court has made such a direction The RICS Register of Expert Witnesses was established for this purpose.

Rule 35.6 enables a party to put questions to an expert instructed by the other party whether the expert is single or joint-appointed. Such questions have to be put within 28 days of the expert's report being submitted. They must be for the purposes only of clarification and thus not seeking further evidence or opinions. The expert's answers will be treated as part of his expert's report. Failure to answer the questions may result in the court making an order that none of his evidence should be relied upon, or that his fees and expenses cannot be recovered from any other parties.

In practice, great care is required to ensure that any questions raised on clarification are in fact just that and not a "fishing expedition" for issues outside the scope of the expert evidence already given. In the case of a single joint-appointed expert, such questions take on even greater importance given that the role must be perceived as wholly even-handed. Practice directions require that a written question to an expert must be sent in addition to the solicitor acting for the party for whom the evidence was given.

It should be noted that, in such circumstances, the court can issue directions in regard to the expert's fees and expenses and limit the amount that is paid. The instructing parties are jointly and severally liable for the payment of the expert's fees. There are pitfalls for the unwary in regard to such an appointment and this will be dealt with more comprehensively in chapter 13.

The assessor

Under Rule 35.15, the court has the ability to appoint an assessor to assist the court in dealing with a matter in which the assessor has skill and experience. The role must be distinguished from that of being a single joint-appointed expert. Essentially, the assessor is an adviser to the judge or arbitrator. He may, however, be asked to prepare a report for the court on any particular issue and may attend the whole or any part of the trial or hearing. However, if a report is prepared for the court, a copy must be sent to each of the parties and may be used by the parties at trial. Practice direction 6.4 provides that while the report is available to the parties, the assessor will not give oral evidence and will not be open to cross-examination or questions.

The position on assessors is similar to that adopted, in my experience, in arbitrations. I have frequently appointed a legal or other technical assessor to assist me having given requisite notice within directions to both parties. The assessor's report (unless they agree to dispense with the requirement) is supplied to both parties and they may comment upon it. The assessor will not usually give evidence orally or be cross-examined. Also, he would not have to respond necessarily to any written representations made, unless the parties have agreed otherwise in which case the assessor will have been aware of the extended role from the outset.

There will be occasions when there is difficulty in obtaining access to information that is relevant or likely to be relevant in the preparation of expert evidence on a particular issue and which is only available to the other party. In some cases, unfortunately, it may become quite obvious that obstructive tactics are being employed outside the spirit of civil litigation within the CPR.

Rule 35.9 disclosure

Fortunately, Rule 35.9 provides possible remedies where a party has access to information that is not reasonably available to the other party. The court may direct the party to prepare a file or document recording the information and serve it on the other party. In my experience, experts are either unaware or loathe to ask their instructing solicitor to make such an application, but certainly this is an avenue that should be explored with the instructing solicitor where such information is not forthcoming.

An example of this was a case where I was involved as expert and there was a confidentiality agreement affecting a particular transaction. The information that was held by the other party was not released. The mere threat of an application within Rule 35.9 was sufficient to deliver the information. A similar situation often happens in arbitration and it is usually sufficient for the arbitrator to give consent for an application to be made to the court for an order of disclosure, resulting in the information being provided (see reference to *South Tyneside Borough Council* v *Wickes Building Supplies Ltd* [2004] NPC 164 on p 134).

Timetables

Practice directions give overall guidelines for a typical case timetable. In practice, in the case of fast-track cases, expert evidence would have to be exchanged within around 14 weeks following directions. However, in my experience of multi-track cases in the High Court involving the more complex disputes, the time limits may be extended, sometimes quite substantially, to allow for the various stages of disclosure and exchange of evidence to be undertaken. It is of the essence of multi-track that there is some flexibility.

The court will arrange a case management conference, skeletal "case summaries" having already been filed with the court. The court directions will thereafter follow, dealing with the disclosure, exchange of witness report and expert evidence, the court will provide for a series of "milestones" when it will review progress and, if necessary, make further directions.

Disclosure

The CPR have completely overhauled the old rules of discovery and renamed the process as disclosure. Since an expert witness may wish to rely upon documents obtained from the other party in order to prepare an expert opinion, it is important to have some understanding of the degree of disclosure, which is now only permitted in respect of four types of document:

1. Those upon which the party relies in support of his case.
2. Those that adversely affect the parties' own case or support the other party's case.
3. Documents not falling within one and two but which are nevertheless relevant and necessary as part of the "story" or "background".
4. The class of documents called "Peruvian Guano", named after that action, which do not prove or disprove the matter initially but could lead to a train of enquiry that might help to advance the case or damage the case of the other side.

Categories 1 and 2 above, known as "standard disclosure", are usually the only ones allowed in small claims and fast-track cases. In multi-track cases, items within categories 3 and 4, may be additionally directed for disclosure. That, however, will be very much the exception rather than the rule.

There are rules defining the nature of the document and scope of the duty to comply with an order for disclosure, which is outside the scope of this book.

Pre-action protocols

In this brief overview of CPR, mention must be made of pre-action protocols. The guidance given may impact upon the expert, particularly where he is asked to advise in the early stages before a claim is actually made. The 1996 Access to Justice report of Lord Woolf recommended the development of pre-action protocols "to build on and increase the benefits of early but well-informed settlement which generally satisfy both parties to the dispute".

The aim of protocols, therefore, is to encourage more contact between the parties at an early stage prior to action, the exchange of information and investigations by both sides, thus enabling the parties to be in a position to possibly settle cases without litigation. It will also enable proceedings to run more easily within subsequent court timetables if litigation does subsequently become necessary.

Accordingly, a number of protocols have been approved regulating pre-action conduct by the parties. It is worth noting that if the court subsequently decides that there has been significant non-compliance with the relevant protocol, it could impose sanctions and/or an adverse order for costs. General guidance is given on the scope of the protocol and the general procedures to be followed, including the format of letters and response to be made by the defendant, including time limits.

Protocols include the provision for notifying the other party in regard to the obtaining of expert evidence. Pre-action protocols have not yet been issued in respect of all types of disputes but the following can be mentioned since a valuation expert may be involved on quantum issues:

- Construction and engineering disputes.
- Professional negligence.
- Dilapidations — in use but awaiting formal adoption.

The full text of CPR Part 35 and the Practice Direction (PD 35) can be found in appendix 1.

Code of Guidance on Expert Evidence The Civil Justice Protocol

The advent of the CPR has spawned a number of codes of practice for expert witnesses and those that instruct them and which, unfortunately, have been the source of some confusion over compliance.

Both the Academy of Experts and the Head of Civil Justice (working with the Expert Witness Institute) have produced Codes of Practice. The latter document was issued in December 2001 and was known as the *Code of Guidance*. Immediately prior to the publication of this book, the

Code of Guidance was replaced by a *Protocol for the Instruction of Experts to give Evidence in Civil Claims*, which has been drafted by the Civil Justice Council with the approval of the Master of the Rolls.

It is perhaps a little unfortunate that it is has been described as a "protocol". It should not be confused with the pre-action protocols described above. It gives advice to experts and those who instruct them "as to what they are expected to do in civil proceedings". Issued in June 2005, it will be effective from 5 September 2005.

The protocol generally follows and amplifies the CPR and is mandatory in some respects. In the preparation of the protocol, the authors acknowledge the assistance received from both the Academy of Experts and the Expert Witness Institute. I will refer to it throughout this book as the Civil Justice Council Protocol, the full text of which can be found in appendix 2. I will not repeat the protocol verbatim but summarise the main points below.

Scope of protocol

The protocol offers guidance to experts and those instructing them. It is intended to assist in the interpretation of Part 35 of the CPR and the associated Practice Directions (PD 35). It does not, however, remove the need for experts to be familiar with them.

Pre-action

Given the current emphasis in litigation of encouraging early resolution of claims out of litigation, it is no surprise that this emphasis is repeated. The early exchange of full information about expert issues in a prospective legal claim are encouraged, enabling the parties to avoid or reduce the scope of litigation by agreeing the whole or part of the expert issues.

Application

The protocol applies to all experts working within the civil actions covered by the CPR but subject to any "specialist" proceedings where other court rules may apply. It is outside the scope of this book to investigate these. An expert should always check with those instructing him at the outset in regard to any specialist protocols or rules that may apply to the particular claim. It is important to note that "the courts may take into account any failure to comply with this protocol when making orders in relation to costs".

Duties of experts

While acknowledging the duty to exercise reasonable skill and care to those instructing them (and compliance with professional codes of ethics), it is emphasised that CPR 35.3 imposes an overriding obligation to help the court and this takes precedence.

Experts are therefore bound by the overriding objectives to deal with cases "justly", "fairly", "expeditiously" and "proportionately", as already described in chapter 3. However, this does not extend to acting as a mediator or deciding any issues that are for the court.

Experts are to provide independent opinions only within their expertise. A suitable test being that the same opinions would be provided for the opposing party on the same instructions. Experts must not indulge in advocacy.

As a result of the judgment in *Phillips* v *Symes* [2004] EWHC 2330, experts are warned that any breach of duty to the court resulting in significant expense being incurred may result in an order for wasted costs against the expert. Further reference to this case and the possible exposure to wasted costs is referred to in chapter 14 in regard to expert witness liability.

Expert as adviser

Part 35 of the CPR only applies to evidence to be relied upon by the court. Other advice remains confidential. Thus, the protocol does not apply where the surveyor is acting as an adviser, but this may change if subsequently such advice transforms into expert evidence or is disclosed as such.

Appointment of experts

As might be expected, the terms of appointment should be agreed at the outset. Guidance only is then given as to what might be included:

• Capacity (eg, a party appointment expert or SJE or adviser).

- Services required.
- Date for delivery of report.
- Basis of expert's charges (including time required).
- Cancellation charges.
- Fees for court attendance.
- Payment.
- Fees payable to a third party.

I provide comprehensive advice on what should appear in conditions of engagement in chapter 7. However, I am not sure whether some of the above points are particularly relevant and some important items appear to be excluded.

Payment time is not often included in conditions of engagement, but it is usually prudent to incorporate this when confirming the billing regime to be followed (see chapter 7).

Fees

It is, once again, confirmed that fees must not be contingent upon the outcome of litigation since this would compromise independence.

Instructions

The scope of instructions required should be noted. It would be advisable to tactfully draw attention to any omissions when receiving instructions and seek clarification.

Acceptance of instructions

It is important that the expert confirms without delay if the instructions appear to fall outside possible compliance. In practice, this is a common difficulty often due to the lack of technical understanding by those instructing. The protocol mentions some possible examples:

- Work required outside expertise.
- Instructions insufficient.
- Cannot fulfil other terms of the appointment.
- Work required that would involve conflicts with expert duties, ie within CPR.
- Inability to comply with any orders made by the court.

Experts must not express an opinion outside the scope of their expertise (or accept instructions to do so).

Withdrawal

The expert is advised to discuss his intention to withdraw with those instructing him and then give formal written notice if required.

Expert's right to ask for directions

Within CPR 35.14, the expert has the ability to apply to the court for directions. It is important to have a prior discussion with those instructing the expert since, obviously, any such application would be a very radical action in the context of the solicitor/expert relationship. In extremis, however, given the difficult circumstances that, in my experience, can occasionally arise, it may be a useful reminder at least of the expert's powers. This issue is referred to again when dealing with the ongoing appointment in chapter 7. The protocol provides the details that must be supplied in any application to the court for directions.

Providing information

The powers contained in CPR 35.9 to ask the court for disclosure of information may be useful where information that is required for the preparation of an expert opinion is not forthcoming from the other side. However, it should be noted that, except where a document is essential, the expert should assess the cost and time involved and whether its provision will be proportionate in the context of the case.

Contents of expert's reports

This is a further application of the rules in CPR 35 and PD 35 flowing from the general obligations and the overriding duty to the court. There is extensive additional guidance that may be summarised as follows:

* Contents governed by scope of instructions, CPR 35 and PD 35.
* Maintain professional objectivity and impartiality.
* Report to be addressed to the court PD 35 (paragraph 2).
* Model forms of experts' reports available from the Academy of Experts and the Expert Witness Institute.

- Must include the Statement of Truth. The form of the statement is as follows:

  I confirm that insofar as the facts stated in my report are within my own knowledge I have made clear which they are and I believe them to be true, and that the opinions I have expressed represent my true and complete professional opinion.

  This wording is mandatory and must not be modified.

- Qualifications of the expert to be included commensurate with the nature and complexity of the case.
- Where reports or opinions are relied on, details must be given, including the qualifications of the originator.
- Keep fact and opinion separate.
- Distinguish clearly between facts that experts know to be true and those facts that are assumed.
- Where material facts are in dispute, experts should include separate opinions on each hypothesis.
- Views should not be expressed in favour of one or other disputed versions of the facts unless the set of facts is improbable or less probable. If so, give reasons.
- If using published sources, state the qualifications of the originator.
- Where there is no available source for the range of opinion, the experts may express opinion on what they believe to be the range that other experts would arrive at if asked.
- Statement of the substance of all material instructions should not be incomplete or otherwise tend to mislead. Instructions should include all material placed in front of experts in order to gain advice.

Conclusions

This is mandatory. It should be at the end of the report after all reasoning. The benefit to the court may be heightened by including a short summary at the beginning of the report (see chapter 11 for my recommendations in this regard).

Amendment of reports

I deal with this issue at some length in chapter 11. Given the practice of some instructing solicitors and/or counsel historically to seek substantial amendments to experts' reports, it is worth repeating the CPR requirements that:

> Experts should not be asked to, and should not, amend, expand or alter any parts of reports in a manner which distorts their true opinion, but may be invited to amend or expand reports to ensure accuracy, internal consistency, completeness and relevance to issues and clarity.

Experts should generally follow the recommendations of a solicitor in regard to the form of reports but maintain independent views as to opinions and the contents expressed, excluding any suggestions that do not accord with their views.

Where an opinion is subsequently changed, a signed and dated addendum memorandum to that effect is generally sufficient. However, in some cases the court may benefit from having a completely amended report (where the cost is justified). Where an expert's report is significantly changed, the report should be amended and should include reasons for the changes.

Any amended report should be clearly marked with the changes.

Written questions to experts

The procedures to be followed are set out and the sanctions that may be imposed if there is a failure to do so.

Single Joint Experts (SJE)

This further expands the procedures where an SJE is instructed, including responsibility for fees (jointly and severally liable). Chapter 13 provides more detailed advice on such appointments.

Discussions between experts

This is an area much fraught with difficulty in practice and I deal with the issues involved in chapter 9. The main points in the protocol arising from CPR can be summarised as below:

- Arrangements for experts' meetings to be "proportionate" to the value of cases. Practicalities may be limited to telephone or video conferencing.

- Parties, lawyers and experts should co-operate in producing an agenda.
- An agenda should indicate agreed matters and summarise those that are at issue.
- Experts must not be instructed to avoid reaching agreement on any matter within the expert's competence.
- Lawyers may only be present if all parties agree (or court orders). If they attend, there must be no intervention unless to answer experts' questions.
- Contents of discussion between experts should not to be referred to at trial unless the parties agree (CPR 35.12 (4)). See chapter 15 for advice on the protection afforded by the privilege rules.
- A statement of agreed and unagreed matters to be prepared and signed by all parties plus a list of all other issues discussed not in the agenda and recommended further action.
- Agreements do not bind the parties unless they agree (CPR 35.12 (5)).

Attendance of experts at court

- There is an obligation to attend if called upon. The experts must take reasonable steps to be available.
- Those instructing experts must ascertain availability and keep experts informed of timetables and the location of the court.
- Consideration must be given to providing evidence via video link. Experts must be informed if trial dates are vacated.
- Experts would normally attend without the need for a witness summons. However, any use of a summons does not affect the rights to pay experts' fees.

Valuation surveyor experts will need to follow this protocol, which is operative from 5 September 2005, as an adjunct to the CPR and the CPR Practice Directions. Where the mandatory parts of the protocol appear to be at all in conflict with the *RICS Practice Statement and Guidance Note* referred to below, then the protocol should be followed at least as far as civil actions are concerned.

RICS Practice Statement and Guidance Note (2nd ed)

To its lasting credit, the RICS published the first edition of the *Surveyors Acting as Expert Witnesses RICS Practice Statement and Guidance Note* in 1997, well in advance of the date when the CPR came into force, which was 26 April 1999. However, it did post-date Lord Woolf's annual report in July 1996, by which time the future requirements and rationale of the CPR were well known.

The *Practice Statement* is mandatory for all chartered surveyors required to provide expert evidence to what is called a "judicial body". RICS Practice Statement byelaw 19(5) and Conduct Regulations make it the duty of all members to comply. It often comes as a surprise that it applies in all cases where expert evidence may be relied upon by any judicial or quasi-judicial body in the UK (PS1.1). It is made clear, for the avoidance of doubt, that such bodies include courts, tribunals, committees, inspectors, adjudicators, arbitrators and independent experts (but not mediators).

There has been much debate in the profession as to whether it should really have applied in the case of evidence given to independent experts. In my opinion, an independent expert, although required to reach his own valuation opinion, will nevertheless wish to have the ability to view evidence put to him as being reliable and wholly unbiased, honest and truthful as if he were an arbitrator. There is, therefore, no reason why there should be any distinction in the case of opinion evidence to an expert in a submission. As an independent expert, I find this requirement to be most helpful in practice.

The *Practice Statement and Guidance Note* has been revised in the second edition as from 2001 and now fully takes into account the updated CPR and compliance with them. There is specific reference to the possibility of disciplinary measures being taken if the expert does not adhere to the *Practice Statement*.

The expert evidence provided by chartered surveyors must be, and must be seen to be, the independent product of the surveyor. The surveyor must also believe that the facts upon which he relies are complete, true and that his opinions are "correct".

Overriding duty

The primary and overriding duty of the surveyor is to the judicial body to whom the evidence is given (PS2.1).

Independent, objective and unbiased opinion

The duties follow very much the CPR principles, including the duty to be truthful as to fact, honest and correct as to opinion and complete as to coverage of relevant matters. The evidence must be independent, objective and unbiased (PS2.2 and 2.3). It applies equally to written representations or to a report where oral evidence is to be given. It emphasises the need to have the requisite knowledge, experience and qualifications and training appropriate for the assignment and resources to complete within timescales (PS3.1).

Taking instructions

There are particular provisions in regard to the way in which instructions are accepted, including the stipulation that the *Practice Statement* and, if appropriate, CPR will apply. The surveyor being instructed must first offer to supply a copy of the *Practice Statement* upon request and confirm that the overriding duty of the expert is to the judicial body (PS3.2). There must be written terms of engagement covering all the matters referred to.

Conflicts of interest

There is reference to conflicts of interest, including any potential conflict that may arise after instructions have been accepted. These must be notified immediately. In line with the relevant CPR provisions, PS5.1 requires that an expert's report must (unless otherwise directed) be addressed to the "judicial body".

Substance of instructions and statement of truth

The surveyor must state the substance of material instructions, the basis upon which the report is prepared, and verify with a statement of truth at the end of the written evidence (PS5.3). The wording to use is as follows:

> I confirm that insofar as the facts stated in my report are within my own knowledge I have made clear which they are and I believe them to be true and that the opinions I have expressed represent my true and complete opinion.

Report declarations

PS5.3 provides that the report must be personally signed and dated. Further declarations to be included at the end of the written evidence are as follows:

- That the report includes all facts that a surveyor regards as being relevant to the opinion that he has expressed and that the judicial bodies' attention has been drawn to any matter that would affect the validity of that opinion.
- That the report complies with the requirements of the Royal Institution of Chartered Surveyors as set down in the *Surveyors Acting as Expert Witness Practice Statement*.
- That the expert understands his duty to the judicial body and has complied with that duty.

Advocacy and expert evidence

Where the surveyor assumes the role of advocate, this must be clearly distinguished from the production of expert evidence and declared to the judicial body and the opposing party (PS9.1). Thus, as highlighted earlier, where a surveyor represents a client in a rating appeal before a local valuation tribunal, he should clearly distinguish when he is arguing the case and when he is providing his expert opinion.

From my own experience of appearing before local valuation courts (as they were previously called), this is not a solution that it is easy to activate in practice. It often proves difficult to disengage from the general thrust of urging the tribunal to accept a particular basis of assessment in order to provide an objective valuation opinion. However, the alternative of making no delineation can, in my experience, prove to be even more hazardous and lead to a blurring of the issues, uncertainty, confusion and therefore lack of credibility.

As a chairman of a valuation tribunal, I normally find no difficulty in separately weighing up the two distinct parts of an oral submission as long as there is no inconsistency or conflict. However, in my experience, it is rare that the *Practice Statement* is fully adhered to in this respect.

Guidance Note

The *Guidance Note* is just that, but should normally be followed since the actual provision is wholly within the spirit of the CPR. In the introduction, the *Ikarian Reefer* judgment is given prominence, re-emphasising that the duty to the client becomes secondary once a surveyor is instructed as an expert to provide evidence for any judicial or "quasi-judicial" body.

Given their importance in terms of "best practice", a brief synopsis is provided of the subjects covered in the *Guidance Note*.

Guidance Note 1.1 — Scope

The *Guidance Note* is for any chartered surveyor who may be required to provide expert evidence in the UK to any judicial or quasi-judicial body, including independent experts. It covers virtually all occasions when a chartered surveyor valuer is required to give expert evidence.

Guidance Note 1.2 — Practice statement

The *Guidance Note* must be considered in conjunction with the *Practice Statement* and provides information on good practice.

Guidance Note 1.3 — Relationship with CPR

The *Guidance Note* is prepared in light of the CPR and comments from the courts on the duties and responsibilities of expert witnesses. There may be additional requirements for expert witnesses where CPR applies, for example giving reasons for the expert's opinion where a range of opinion is possible. Additionally, surveyors must comply with CPR, The Civil Justice Council Protocol and be aware of any changes.

Guidance Note 1.4 — Neutrality

Neutrality is of the utmost importance. The expert's overriding duty is to the judicial body. The *Guidance Note* and *Practice Statement* are intended to ensure impartiality and to prevent the expert acting as an advocate.

Guidance Note 1.5 — Informing the client

There is an obligation to make clients aware when accepting instructions to act as an expert witness of the expert's obligations to make a declaration confirming belief in the accuracy and truth of evidence in reports.

Guidance Note 1.6 — Ikarian Reefer

The judgment of J Cresswell is summarised in chapter 3, pp 11 and 12.

Guidance Note 2.1 — Knowledge and resources

Surveyors must be satisfied that they have the knowledge and resources to fulfil the task adequately within the allocated time span.

Guidance Note 2.3 — Secondary duty to client

The duty to the client is secondary once the surveyor is instructed as an expert to provide evidence for a judicial body. If not acceptable to the client, instructions should be declined.

Guidance Note 2.4 — The various roles of the expert

This guidance note discusses the various roles that may be adopted by an expert, including establishing facts, giving expert opinion and conducting enquiries on behalf of the judicial body.

Guidance Note 3.1/3.2 — Initial advice to the client

Comments upon prior initial advice to a client before being instructed as an expert and the question of whether such advice should be routed through the client's solicitor with the aim of attracting legal or litigation privilege. (I comment on this difficult issue on pp 118–121.)

Guidance Note 3.3 — Advice to client prior to accepting instructions

In my experience, surveyors are not often aware of the requirement to notify in this context. An expert should make the client aware of the *Practice Statement* and offer to supply a copy. It is important to agree terms of engagement and be satisfied that no conflict of interest can arise. If the advice given in chapter 7 in regard to conditions of engagement is followed, these matters will be covered adequately.

Guidance Note 3.4 — Overriding duty

It is re-emphasised that the overriding duty is to the judicial body. This should be made clear to the client and this takes precedence over any duty to the personal body by whom the expert is instructed (Guidance Note 3.5).

Guidance Note 4.1 — Duties of the expert witness

This merely repeats the *Practice Statement,* Guidance Note 4.2 and 4.3 and further expands the philosophy and advise that surveyors should decline instructions if they feel they cannot fulfil the obligations.

Guidance Note 5 — Accepting instructions

This stipulates that the expert must ascertain the *identity* of the parties and check for conflicts of interest that may arise out of a previous involvement with any party, the dispute or the property, which might undermine his independence. The surveyor should either refuse the assignment or seek further instructions following disclosure.

Guidance Note 11 — Oral evidence

This reaffirms that oral evidence must be truthful and the expert's honest opinion. If the expert does not know the answer to the question, he should say so rather than endeavouring to give a possibly misleading or incorrect answer. The answers to questions must be addressed to the judicial body not the advocate posing the questions. A surveyor should never attempt to advocate the merits of his client's case.

Guidance Note 7.2 — Evidence of fact

All facts relevant to the case must be stated truthfully and fully whether or not they favour the appointing party.

Guidance Note 7.3 — Confidentiality agreements

The expert is asked to be aware that evidence to a judicial body takes precedence over any contractual, professional or other duties that may conflict with confidentiality agreements. While it is agreed that these cannot be ignored, this guidance may have to be revised somewhat in light of recent case law (*South Tyneside Borough Council* v *Wickes Building Supplies Ltd* [2004] NPC 164 — see chapter 16 on Arbitration).

Guidance Note 8.1 — Opinion evidence

Opinion evidence must be honest, objective, independent, unbiased and non-partisan. The expert surveyor should not exaggerate or seek to obscure alternative views.

Guidance Note 15.2 — Report cover sheet

It is stipulated that the front sheet of an expert's report should not obscure the name of the expert witness, the proceedings and the nature of the evidence (see appendix 4 for an example).

Guidance Note 15.3 — Form of expert report

There is comprehensive guidance on the format of the expert's report and all of the advice is reflected in chapter 11 on this subject and in my example in appendix 4. The precise format will be flexible and adaptable according to circumstances. The items set out in Guidance Note 15.3 should be taken as the minimum that should be provided in a report arranged normally in the order that is advised.

Guidance Note 17 — Meetings between experts

Some useful guidance is given in regard to meetings between experts and the general aims of narrowing the issues and agreeing facts. Set out below in chapter 9 are general thoughts and advice that complement Guidance Note 17.

Guidance Note 18 — Fees

It is stressed that the expert must set out clearly the basis of his fees and have this agreed at the outset. The importance of keeping time records is also correctly shown having regard to the possibility that the expert's fees will form part of a costs award. This aspect is dealt with more fully in chapter 8.

Guidance Note 19 — Contingency fees

I wholeheartedly agree with the advice that any form of contingency fee is incompatible with the duty of impartiality and independence required by the expert witness. This should be avoided. It may be necessary to advise a client in a pre-existing relationship, for example in the case of a rating appeal or rent review, that the basis of fees will have to change when an expert witness role is assumed.

Guidance Note 22.4 — Expert opinion and advocacy

This contrasts the two roles and the need to differentiate them in practice. Further advice is given in Guidance Note 22.5 that, in order for the two roles not to be confused, advocacy should be undertaken from a different position in a tribunal room (for example). I believe that this is unnecessary and may be unacceptable to an arbitrator or valuation tribunal. A clear division of the presentation will usually suffice.

Guidance Note 22.6 — Advocacy and written submissions

I fully concur with the advice given that in the case of written submissions a solution would be to divide the presentation (ie advocacy) from the expert's reports and submit in separate envelopes. I have yet to receive any submissions where this has been done!

The guidance notes are not mandatory but are intended to embody "best practice". These are procedures that "in the opinion of the RICS meet a high standard of professional competence". There are important implications, therefore, if they are not followed. While public policy dictates that an expert witness has generally immunity from action as a result of his evidence being "wrong", the evidence given by the chartered surveyor expert may nevertheless be challenged by the court, tribunal or arbitrator if there are no substantive reasons for departure. In some circumstances this might give rise to at least a claim for "wasted costs" (see chapter 14) where, as a result, an action has been unnecessarily prolonged or the judicial body has been misled.

The *RICS Practice Statement and Guidance Note* was undergoing revision at the time of publication. The consultation draft of the proposed 3rd edition has been updated and enlarged mainly to clarify issues of great concern to the RICS Dispute Resolution Service and Faculty. For example, it includes specific new references to such matters as the suitability of the expert in light of previous involvement in negotiations, (see chapters 6 and 7). Other significant proposed changes are:

- Where there is a potential conflict of interest, a surveyor is obliged to advise those instructing him in writing that legal advice, as to the advisability of the surveyor's continued appointment, is sought.
- A reminder to surveyors, and those instructing him, that there are limits to the changes that an expert may be asked to make in his report.
- Surveyors are to use all reasonable efforts to agree matters of fact and to refuse instructions where asked to limit agreements between experts.
- The difficulties and conflicts in an advocacy/expert witness role have been expanded upon. The surveyor must reflect fully on the advantages and disadvantages of the dual role.
- One of the most important new proposals is that any form of contingency fee arrangement is specifically outlawed in any expert witness appointment. This will have implications for many rating surveyors and rent review surveyors, whom I suspect do not always make the change, as discussed later in chapter 7.

There are also substantial and wide-ranging additions to the *Guidance Note*, together with some expanded and updated recommendations. These include the inclusion of a chronology of events and an executive summary in an expert's report.

The consultation draft of the proposed 3rd edition can be obtained from RICS Business Services Ltd.

These proposals are generally to be welcomed, particularly in the context of defining the surveyor's expert duty where a negotiating role had been assumed. The specific exclusion of a previous or current contingency fee will do much to underline the wholly different nature of the expert role as opposed to advocate or negotiator.

Civil Evidence Act 1995

Valuers, whether operating in the residential or commercial property markets, will be aware that, in the majority of cases, comparable transactional evidence upon which they seek to rely in arriving at an opinion of value will derive from transactions in which they have not been personally involved. It is rare for all or a major part of the market evidence to have been dealt with at first hand.

Unproved evidence before 1997

Prior to the Civil Evidence Act 1995, considerable difficulties were experienced in terms of "proving" unagreed evidence of comparable transactions in a property dispute since, under the then court rules, any such secondary evidence being hearsay was wholly inadmissible. Therefore, it was often necessary, where not agreed, to have the evidence proved by attendance at the court by a principal or agent involved in the transaction or other means acceptable to both parties, eg by affidavit. This placed an unnecessary burden upon the pursuit of justice in such proceedings since the reality was (and is) that the valuer necessarily relies on unproved evidence of transactions. In the valuation process, such weight as is believed to be appropriate will attach to such evidence in arriving at an opinion.

New position on hearsay

In civil actions, the position has been corrected by the Civil Evidence Act 1995, which came into force on 31 January 1997.

The Act states that "in civil proceedings, evidence shall not be excluded on the ground that it is hearsay". Civil proceedings is defined as meaning "civil proceedings before any tribunal, in relation to which the strict rules of evidence apply, where there is a matter of law or by agreement with the parties". There is, however, a condition of this relaxation that any party proposing to adduce hearsay evidence must give advance notice to the other party and provide, on request, reasonable and practicable particulars of the evidence. Failure to give such notice and such particulars will not render the evidence inadmissible, but it can be taken into account when considering the weight to be given to the evidence. The other party may, with the court's leave, call a witness and cross-examine him in relation to any hearsay evidence proposed to be adduced.

By virtue of Section 11, the Act applies to civil proceedings before any tribunal in relation to which the strict rules of evidence apply whether as a matter of law or by agreement of the parties.

Admissibility and weight

The upshot of this statutory change is that emphasis has been removed from admissibility of hearsay evidence to the weight that may be attached to it, given all the circumstances, but subject to the requirement to give notice as appropriate. It is further provided that, in estimating the weight (if any) to be given to hearsay evidence, the court shall have regard to any circumstances from which inference can be drawn and will have regard in particular to:

- Whether it would be reasonable and practicable for the party by whom the evidence was adduced to have produced the maker of the original statement as a witness.
- Whether the original statement was made contemporaneously with the occurrence or existence of the matters stated.
- Whether the evidence involves multiple hearsay.
- Whether any person involved had any motive to conceal or misrepresent matters.
- Whether the original statement was an edited account, or was made in collaboration with another or for a particular purpose.
- Whether the circumstances in which the evidence is adduced as hearsay are such as to suggest an attempt to prevent proper evaluation of its weight.

As far as actions in the courts are concerned, under CPR, it is worth having a dialogue with the solicitor instructing you in regard to the inclusion of hearsay evidence in the expert's report to be provided. In practice and in relation to High Court actions as an expert witness, I have not recently encountered any difficulties in relation to this issue since normally both experts have the same degree of difficulty and it is therefore usually accepted. The valuer's opinion, particularly on the more complex issues, will often have to depend upon external evidence, not only in relation to comparable transactions but also other matters, such as planning, building costs and environmental issues.

In relation to evidence before tribunals, the same comments generally apply. Regard should be given to the Lands Tribunal rules but no difficulties are likely to be encountered before local valuation tribunals.

Arbitration directions — evidence

In relation to arbitrations, the arbitrator should have included in his directions precisely how he intends to deal with the proving of evidence. He will normally either direct all hearsay evidence as admissible but will attach a degree of weight, as appropriate, in reaching his award, or he will merely prescribe that the Civil Evidence Act 1995 shall apply with some practical consequences. However, within the provisions of the Arbitration Act 1996, the arbitrator will only be free to direct where a contrary prior written agreement has not been entered into between the parties on procedural issues.

Agreeing facts

In practice, however, any difficulties arising in the course of hearsay evidence will often be circumnavigated by compliance with the directions of the court (resulting from the CPR requirements), arbitrator or tribunal. The parties should prior agree a statement of facts, including as much of the comparables as possible so that questions of hearsay and admissibility are no longer an issue. This will avoid any challenge on comparable transaction details. Experts should ensure that there is provision for an agreement on facts, if this is not already directed by the court or tribunal, if possible, prior to the exchange of experts' reports or the making of submissions.

Prior to the Appointment

6

In this chapter, focus is given to matters that need careful consideration and appropriate action when a new appointment is being contemplated. The Civil Justice Council Protocol enlarges upon the CPR requirements on such matters in the case of civil actions (see chapter 4 and appendix 2).

My comments and advice generally apply whether the appointment is an entirely new matter for the expert valuer or a continuation of an existing instruction. Examples of the latter include a rating appeal adviser or rent review negotiator, where proceedings to a tribunal or arbitrator have become likely because a settlement has not been achieved. In such cases, the circumstances may stimulate different responses.

Suitability and availability

In the case of a new appointment and perhaps, where there is to be a continuing role, three issues need to be immediately considered:

- Whether the valuer has genuine relevant expertise and experience to give the evidence required.
- Is the expert truly independent? Are there conflicts of interests?
- Is the valuer likely to be available to undertake the work in accordance with the directed timetable within the context of "milestone" dates, including a trial hearing? Is it possible to comply with any other timetables, eg the date of response to the claim or within pre-action protocols?

Test of genuine expertise

The question of genuine expertise has been referred to earlier in the context of court rules and codes of practice. This is usually obvious but, if there is any doubt, err on the side of caution. Do not take the instruction. In some circumstances, it may be helpful to apply the tests referred to earlier. Is the valuer someone who would be viewed in the wider commercial market as an expert by potential clients and would he be someone to whom a case would be referred to with confidence by his peers for a second opinion?

The expert must be involved on a day-to-day basis either as a consultant or agent within the particular property market or markets. It is equally important that the expert is of some standing within his professional field. It is often helpful to have had some experience of litigation and appearances before courts or tribunals but, of course, the lesser-experienced, younger valuer will have to start somewhere! In the latter case, it may be reasonable to say that his views have been "tested" by a more experienced colleague.

Conflicts

One of the first tasks to be confronted with a possible new appointment is the conflict of interest search. Those wishing to instruct the expert will supply the names of the parties and the subject property or properties. It is important that you ensure you have the names of the appropriate parent companies, associated companies and major subsidiary companies of the parties in order to make the widest possible check. Opinions vary as to the degree to which a conflict of interest check is necessary in the case of an expert witness. Given that the expert's duty is to the court or tribunal, it might be argued that since this takes precedence, other relationships are less important. However, this is unlikely to be the view of the other party, who may have justifiable objections as to independence, impartiality and thus credibility. This may influence the court or tribunal etc when attaching weight (or any weight) to the evidence presented.

In practice, the party on whose behalf the expert evidence is to be given will be greatly concerned with any relationship with the opposing party. There would be serious doubts concerning impartiality and independence if there have been any close past professional, financial or personal relationships with the company or individual on whose behalf the evidence is to be given. If there are any past or current relationships with you or any member of your practice, then disclosure must be made at the outset. This includes any current situations that could potentially evolve into a professional or financial relationship.

The question of remoteness is a difficult one but it is usually obvious when a test is applied as to whether an involvement, in the eyes of the court or another party, might give reasonably grounds for believing that impartiality or bias might occur. It is a question of the degree of past or current involvement (or the potential for the same) but, if in doubt, disclose. It is not unknown that following disclosure the parties agree in all the circumstances that the expert may still proceed, as it would not be perceived (by the judical body or the parties) as likely to prejudice independence or impartiality given the particular circumstances of the case.

Interestingly, Mr Justice Neuberger in *Liverpool Roman Catholic Archdiocese and Trustees Inc* v *Goldberg* [2001] Lloyd's Rep 518 concluded that an expert giving evidence on behalf of a friend and professional colleague of long-standing did not mean, as a matter of law or even as a matter of fact, that he was incapable of fulfilling the role within the *Ikarian Reefer* principles (they were in the same set of barristers).

However, in *Hussein* v *William Hill Group* [2004] EWHC 208 two friends of the claimant gave medical reports in support of a personal injury claim. The judge ruled that the claimant had exaggerated his injuries and criticised the experts. Their conduct was reported to the General Medical Council. My advice is always to disclose, not least since the judge hinted in the *Goldberg* case that there was "an inbuilt disadvantage, which might attract fertile cross-examination".

Chinese walls

The parties may consider formal "Chinese wall" arrangements. However, a recent Court of Appeal decision in *Marks & Spencer plc* v *Freshfields Bruckhaus Deringer* (2004) 148 SJLB 788, concerned conflicts of interest in relation to corporate bid advisers. It may be of relevance to note that the court rejected the "Chinese wall" concept as being inadequate, going forward, where it had not been previously implemented, ie the stable door was being shut after the horse could have bolted.

Employee as expert for employer

The landmark case of *Field* v *Leeds City Council* [2000] 1 EGLR 54 concerned the much-debated issue of whether an employed person can give truly impartial and independent evidence on behalf of his employer. The Court of Appeal was not prepared to accept that a surveyor employed by a local council was automatically disqualified from giving evidence but, in effect, the fact of the employment may limit the weight that is attached. It was important that a party wishing to involve an employee as an expert should inform the other party and the court of the intention to seek to satisfy the court at an early stage that such an expert could act objectively. This should depend upon whether:

- it can be demonstrated whether that person has relevant expertise in an area at issue in the case
- it can be demonstrated that he is aware of his primary duty to the court if he gives expert evidence.

Given the remaining objective risks, my advice is that employed persons should avoid, if possible, assuming the role of an expert unless there are particularly good reasons (eg a low-value claim) or special circumstances (eg "unique" knowledge). It would be more prudent to instruct a wholly independent external expert.

Deadlines

It is self-evident that, before confirming instructions, the potential expert must carefully consider any immediate or long-dated deadlines for various stages of the work depending upon the circumstances of the case and the position reached. If it is in the pre-action protocol stage, it will be necessary to discuss what deadlines have been agreed between the parties in terms of exchanging information and any initial reports by the experts. Where a claim has already been made and a response is required under CPR, it may be necessary to provide initial advice as a matter of urgency. If the case has proceeded beyond the first case management conference, it may be necessary to consider the directions and the timetable of dates agreed, which will usually be for:

- exchange of witnesses' statements of fact
- exchange of experts' reports

- a meeting of experts if to be undertaken
- a statement of agreed and unagreed facts to be prepared by the experts
- trial.

Be wary of a late call on a Friday evening from the desperate solicitor who has neglected to appoint an expert in accordance with court directions (usually to save costs) and needs urgent action. The fee arrangements offered may sound wonderful but, from bitter past experience, this can be a recipe for near disaster since rarely, in my experience, is the case as straightforward as it first appears. A solicitor who has mismanaged his client's case may be rather "economic with the truth" and the extent of the brief.

Also, from experience, be extremely circumspect when accepting any early initial retainers to act as expert, without thereafter carefully monitoring progress (or the lack of it). Quite out of the blue, the expert may suddenly receive news from the instructing solicitor to the effect that after all a settlement was not achieved and could he have the expert's report by next Friday? (When, of course, the expert had planned to be on holiday!) Some clients — and even members of the legal profession — still labour under the misapprehension that a retrospective valuation of, for example, a multi-let office investment, can be quickly resolved by a mere drive past!

Existing instructions — the changed relationships

It is appreciated that much of the above advice will not be wholly applicable in the case of long-standing instructions on a rent review or rating appeal. Nevertheless, given the changed duties and priorities, it is important to review the relationship with the client.

The valuer is to become an expert before a tribunal/arbitrator (with a primary duty to that body or person) as opposed to acting as an adviser/negotiator. The changed relationship should be emphasised with the client and the fee arrangements adjusted if necessary and, it is suggested, new conditions of engagement entered into.

In most cases, it will be unnecessary to review potential conflicts of interest unless, of course, the client has been prepared to live with your lack of independence, which, nevertheless, might prejudice any evidence given before a tribunal or to an arbitrator.

It cannot be emphasised too strongly that there is the need to educate the client if necessary about the change of roles. The client's best interests will be best served by someone who is not only independent but also seen to be independent and providing honest, objective opinion in accordance with the court rules, the Civil Justice Council Protocol and possibly the RICS practice statement. The dangers of not so doing should be stressed.

The adviser role

Finally, in the case of providing expert evidence in the civil courts, it is necessary to thoroughly discuss, at the earliest stage, the precise context in which any initial advice is to be given, in advance of any expert's report. Will the role be to act as an adviser with the possibility of contributing to, or advising at, mediation?

Availability — the sanction of the courts

In relation to case management directions and timetables, it is vital that full disclosure of any dates of non-availability are given to the instructing solicitor at the earliest opportunity. The Court of Appeal has determined that the non-availability of an expert will rarely provide justifiable reasons for altering case management directions. It is clear from the judgment in *Matthews* v *Tarmac Bricks and Tiles Ltd* (2000) 54 BMLR 139 that it is not sufficient at a case management hearing, when setting a trial date, to have the dates when the expert witness is unavailable. It is necessary to discuss expert witnesses' commitments and the extent to which those commitments can be moved, if necessary. Therefore, the expert witness has an important role in contributing to the potential timetable.

In my experience, the courts generally allow a certain degree of flexibility. But in case of difficulty, they have the ultimate sanction with a hard line being taken if the difficulties of setting dates convenient to all appear to be insurmountable.

On Being Appointed

Existing instructions

The appointment, whether explicitly made or not, of the valuer as expert, for example, in the case of a rent review or rating appeal, may merely be a further stage in ongoing instructions, perhaps after the passage of some months, or even years in the case of a rating appeal. Nevertheless, it is important to recognise that a wholly changed relationship has taken place. In such circumstances, it is strongly recommend that new conditions of engagement are entered into. This may only be in the form of updated advice to the client since a settlement cannot be reached within the terms recommended, it is necessary to proceed to arbitration or to present a case before the tribunal.

There will be cases where an ongoing consultancy may ultimately develop into a situation where expert evidence before a court is required. Where the appointment as an expert witness evolves in this way, it is important that any potential conflicts arising out of the previous relationship as adviser to the client are examined and, if necessary, disclosed, since independence or perceived impartiality may be an issue.

It is suggested that an update letter on the continuing instruction could emphasise:

- The new role that is being undertaken, where the primary duty is to the court, tribunal or arbitrator. The need for compliance with the *RICS Practice Statement* and/or court rules, which will involve giving opinion in a wholly independent and unbiased fashion (in accordance with the *Guidance Note*, it is necessary to offer a copy if required). The dangers of non-compliance should be stressed.

- The reconfirmed fee arrangements will offer a new basis to comply with the *Practice Statement*.

Contingency fee arrangements

I believe it is widely understood that the expert witness must not be "interested" in the outcome of the dispute in respect of which he is to give evidence since this would be incompatible with his duty of impartiality and independence. Therefore, it follows that previous success-related or what are often termed "contingency" fee arrangements will need to be revised to a fixed sum or time basis as appropriate. This is a specific requirement of both the *RICS Practice Statement* and the Civil Justice Council Protocol (see chapter 4).

I suspect that many surveyors when confronted with this difficulty tend to "overlook" the requirements to be bound by a non-contingency fee basis. This may be in the mistaken belief that it will not affect impartiality or that it will not become an issue.

The latter reasoning may be the reality in the majority of cases. However, the valuer should be left in no doubt as to the perils. The expert, in those circumstances, is not only in breach of the guidance notes (and therefore open to disciplinary proceedings if he is a chartered surveyor) but, more importantly, if forced to disclose before an arbitrator or tribunal, this could have a damaging effect on the valuer's credibility as an expert. This might be particularly unfortunate if, in fact, the evidence was truly impartial and independent but becomes stained irrevocably by a success-related fee disclosure. In some circumstances, it might just tip the balance, so that the tribunal or abitrator draws an adverse inference, preferring the evidence of the other side.

I suspect that there still needs to be a major culture evolution in the context of the changed relationships. Perhaps the lessons learned from a bad experience at the hands of an arbitrator or tribunal may hasten the conversion! However, in my opinion, it is better to be forewarned and forearmed.

New appointments — conditions of engagement

I am not an advocate of bureaucratic overkill when it comes to drafting documents for professional service contracts. In my experience, they are often too lengthy and contain much that is irrelevant and superfluous. However, there are exceptions, and this is one of them.

The relationship between litigator and expert witness can be somewhat ambivalent, even under the most sublime conditions. Agendas are different, and expectations may differ. There can be inbuilt tension given the different agendas and duties. All the more reason, therefore, to adequately clarify what is going to be delivered, when, in what form and for how much.

It is not necessary to have a lengthy contract. The extent of it will, to some extent, depend on whether comprehensive instructions were given at the outset, since a good litigator will have covered many of the points. For the purposes of illustration, it is assumed that the litigator has not and the expert being appointed has to fill in many of the gaps. In the case of civil actions, it is worth bearing in mind the specific guidance given in the Civil Justice Council Protocol on terms of appointment (see chapter 4). As a minimum, it is recommended that conditions of engagement should include:

- **Name of action and capacity**
  (If commenced) the name of the parties and court reference (if relevant). Whether appointed by a party as an expert, or SJE or adviser (or possibly both).

- **Case description, background and history**
  A brief reference (if not already provided in the instructions) to the substance of the action circumstances.

- **Identity of party instructing**
  For example, instructions to provide valuation expert evidence on behalf of the claimant.

- **Summary of instructions/services required**
  This should include reference to the description, location and title of the property interest or interests to be valued. The basis and method of valuation, valuation dates, special assumptions or directions, including any reference to relevant law or particular circumstances.

- **Compliance**
  If not already provided, full acknowledgement and confirmation of compliance (as appropriate) with CPR Part 35, Civil Justice Council Protocol, and *RICS Practice Statement and Guidance Note* (2nd ed). An offer should be made to supply a copy of the *RICS Practice Statement* and the opportunity should be taken to reiterate the general principles of independence, duty to the court and confirm declarations, which will be made in the expert's report.

- **Conflicts**
  The issue of conflicts will have been covered prior to the appointment, but it is important to confirm where disclosure has been made and that this has been accepted by the parties.

- **Valuation basis and compliance**
  In the case of providing capital valuations and possibly rental valuations, it is important to confirm, if appropriate, that the valuations will be undertaken in accordance with *RICS Appraisal and Valuation Standards* (5th ed) ("the Red Book") or other established basis if appropriate, eg international valuation codes of practice.

- **Confidentiality**
  This includes confidentiality of any involvement in the litigation prior to and outside of any court hearing (which is, of course, in the public domain). It is important to recognise that any documentary or other details are wholly confidential and must be kept so.

- **Timetable**
  If the action has already commenced and there has been a case management conference, it is likely that a court order has been issued with a timetable for exchange of experts' reports, meetings of experts, etc and these dates should be inserted. Any other pre-action or pre-action protocol timetable agreed between the parties should also be acknowledged. The date for delivery of the report should be confirmed.

- **Availability**
  Any limitations on availability in terms of complying with court timetables or the dates for a trial or hearing should be inserted.

- **Documents and disclosure**
  To the extent that there is any understanding at that stage of the subject action, the opportunity should be taken to survey the availability of case documents from the client's files or disclosed by the other side, together with any witness statement of facts, advice and reports otherwise relevant to the case. In many instances, the expert valuer will have received copies of case documents with instructions but may have to make arrangements to view others or obtain further copies at a later stage.

- **Expert's limitations**
  If not already agreed, this is an opportunity to confirm the extent of the expertise in relation to the advice or expert evidence to be provided. In the event of such limitations, it may be necessary to agree as to other expert evidence required in support, eg building cost inputs in a development valuation.

- **Other experts**
  This concerns the scope of the evidence to be given by other experts that may have a bearing upon a relationship to the valuation evidence or advice to be given.

- **Adviser role**
  It will be invariably the case, particularly in larger actions, that the expert witness will initially act, to some degree, as an advisor to the litigation team up to and maybe beyond any attempts to settle by mediation or whatever. The understanding as to the possible duality of roles should be fully set out with reference made, if necessary, to questions of disclosure (to the court) and any special stipulation on how the dual roles are to be perceived and acted upon.

- **The inspections**
  Whether an inspection is likely to be undertaken early or late, this may be a good opportunity to confirm what is required and to make the assumption that there are no limitations on access.

- **Fees**
  It is likely that fee arrangements would have already been discussed prior to being appointed but the opportunity should be taken to confirm these arrangements either on a fixed fee for

various stages of the work or on a time basis and any differential rates, eg higher scale of time charges for court appearances. It is important to establish the billing regime that may be direct to the client (as opposed to his legal advisor) or aggregated with the solicitor's own fees. It is prudent to specify whether monthly or quarterly and to include reasonable expenses and VAT.

If there is to be direct invoicing to the client and not the solicitor, it may be necessary to make provision for a direct written undertaking so as to ensure privity of contract. Where it is agreed that the valuer expert will have a valuation team in support, probably at lower time rates, this should be confirmed as part of the fee arrangements. Any fee payments to other parties should be specified.

- **Valuation team**
  With larger cases, it will often be necessary to employ a valuation team, perhaps one or two assistants/graduates, to assist in the initial collation of market evidence and other data at lower time rates.

  It is important to confirm these arrangements and to emphasise that while the valuation team will be employed in a supportive role in the initial stages, the selection of market evidence and the preparation of the valuations, inspection of the properties and the preparation of the expert's report will be entirely and solely undertaken by the expert valuer. Only the time-consuming operation of sourcing the basic facts and collecting the evidence together would be subsumed to other members of the valuation team.

- **Type of report**
  In the early stages of litigation, perhaps sometimes even before a claim has been made, a preliminary valuation report may be required but not an expert's report as such. It is important to clarify whether and to what extent a report by the valuer's firm may be required. Such reports will often be used in mediation but it may be necessary, at a later stage, to convert the advice into an expert's report. It may be prudent to seek guidance in regard to the extent to which any initial report may be subject to later disclosure.

In appendix 3, an example set of possible conditions of engagement are given and drafted on the assumption that only a limited letter of instructions was received from a solicitor instructing an expert valuer on behalf of the claimant in a valuation negligence case.

Adviser and expert roles

Although this has been touched upon in connection with conditions of engagement, it would be irresponsible of me not to amplify here given the potential for difficulties and ambivalence. In the years prior to CPR, it was my experience that instructing valuer experts often assumed (or were content not to be dissuaded) that the roles were not divisible notwithstanding the conflict of objectives. Since CPR, and the emphasis on independence and duty to the court, the potential for conflict has been thrown into much sharper focus and, in practice, is a constant source of debate in the expert witness arena.

At the extreme, there can be circumstances where the expert is so much drawn into the adviser role with other members of the client's litigation team in attempts to progress the case to the client's best advantage, that he can no longer view matters with sufficient detachment, independence and objectivity to satisfy his overriding duty to the court as an expert witness. Experience suggests that this will be rare.

It is possible, while exercising proper professional standards and ethics, to undertake the role of the advisor and then, subsequently, revert to the expert role. But only as long as the differences are understood by all concerned and the earlier "advocacy" role is fully protected in law. Thus, it is important that any early advice and possibly settlement discussions in or out of possible mediation are undertaken under privilege, ie on a "without prejudice" basis. As such, any potential conflict in terms of giving honest, objective, unbiased and independent evidence cannot be challenged later.

In my experience, the greater difficulty is how the valuer expert is perceived by those instructing him. The client may need to be reminded from time to time of the shared role. Expectations in regard to the ultimate expert evidence may otherwise become over ambitious or unrealistic as a result of the earlier experience when the valuer was acting as an adviser.

I suggest, therefore, when written or oral advice is given on a substantive point, that the valuer should make it clear in what role he is making the point, with a necessary saving clause or reservation where the ultimate expert opinion could depart. It is not suggested that the expert should ever vary his honest opinion or misrepresent at any stage. Remember that advocates must likewise never distort the facts. However, he may be required to put the best "gloss" that is possible on the facts and be overly persuasive when seeking to achieve a settlement or assisting at a mediation.

If in doubt at any stage how best to proceed, it is best to fully discuss with those instructing you and if the problems of professional ethics are strained to the limit, then it is better that the role is divided, perhaps with a colleague becoming the adviser.

The difficulties of undertaking both roles can lead to considerable dangers, not least the impugning of the expert's integrity if there is long exposure to without prejudice negotiations with the expert on the other side. Although not strictly discloseable in any subsequent court hearing, it is nevertheless available to the other side in terms of their cross-examination tactics of the expert's evidence, which could lead to extreme discomfort or worse!

In a more perfect world, an adviser in a litigation team should never subsequently act as an independent expert with a primary duty to the court. However, this would be impossible without incurring considerable extra costs. It is up to the skill, integrity and ethical sense of the expert/adviser to recognise the limitations and fully discuss with those instructing him how far he can commit to the duality.

In respect of expert evidence arising out of a continuing case where the valuer is already acting as a negotiator, such as in rent reviews or rating appeals, the continuance of a dual role is of the essence. Negotiations on a without prejudice basis will often continue to and possibly beyond the date when written submissions to an arbitrator or expert are required. Maybe even up to the day of the tribunal hearing in the case of a rating appeal. However, this does not, in any way, lessen the need to maintain the ability to successfully undertake the expert witness role if, ultimately, a settlement is not achieved.

In the situation just described, there is a tendency to be over optimistic in the negotiating phase. It is invariably the case that one is more and more persuaded by the merits of one's own advocacy and the arguments deployed with a consequent loss of realism and objectivity. This is a natural tendency for confident and well-motivated professionals.

Therefore, it is suggested that enthusiasm for the case should be balanced in equal measure by an assessment from time to time, quite objectively, of what is likely to be provable before an arbitrator or tribunal after making reasonable allowances for the merits of the other side's agreements. This may involve giving credit for alternative views, which may not accord with one's own. This is not always easy in practice and further illustrates the inbuilt conflict that arises when expert evidence is, in effect, "bolted on" to a previous negotiating role.

It has often appeared to me as arbitrator in rent review disputes that even an experienced valuer will take an unnecessarily jaundiced view of the points put in evidence by the other side. The blinkered approach may well damage credibility.

There is often considerable merit in taking a second opinion from a well-respected colleague, without divulging your own arguments and for whom you are acting in the dispute. I have sometimes been surprised in the past at the outcome!

Valuation compliance

I strongly recommend that any valuation prepared, either capital or rental, for the purposes of an expert's report should comply, if possible, with established codes of practice or guidance notes. In the majority of cases within the UK this will be by reference to the *RICS Appraisal and Valuation Standards* (5th ed) (commonly referred to as the "Red Book"). Compliance is obligatory for chartered surveyor valuers.

The court, arbitrator or tribunal is likely to be more impressed if the leading industry code of practice and standards are adopted. If nothing else, it will lead to some degree of certainty and consistency in terms of definition and applications. Where, for any reason, there is to be a departure from the definitions or non-compliance otherwise with the standards in terms of investigations and procedures, these should be stated.

Directions

I referred earlier to the active case management now required under CPR. Normally, the first task is to allocate the case to small claims, fast or multi-track. This will be determined after the court considers the details supplied in the allocation questionnaire that the parties have to file, a statement of case and defence having been submitted. The allocation questionnaire will include details of how the case will be progressed and any agreement as to case management directions.

If the directions are not agreed, the court will impose its own. If these procedures have already occurred, it is important that the order for directions by the court is disclosed to the expert and any other agreements entered into between the parties in pursuance of the directions made. The court will often undertake to hold a case management conference before directions are actually issued. The directions will normally include:

- Provision for standard disclosure between the parties.
- Disclosure of witness statements by way of simultaneous exchange. Direct that a single joint expert is appointed unless there is good reason not to do so (in the majority of multi-track cases this will not occur).
- Provide for disclosure of experts' reports by way of simultaneous exchange or sequentially where related to quantum of damages.
- Provide for the experts to meet and agree facts, identify issues and reach agreement as far as possible. This will normally include the requirement to provide a statement for the court as to what they agree, what is disagreed and why. In this context, it is interesting to note that any agreement between experts on any issue that does not bind the parties unless they agree otherwise (CPR Part 35.12 (5)).
- A trial period.
- Directions will include dates by which the above are to take place.

Case documents

Where the appointment as expert arises in the case of an action already running and directions have already been made, it can be anticipated that in any multi-track case of consequence a substantial delivery of copy documents easily retrievable from ring-binders will arrive in the expert's office. These normally include:

- Copies of the parties' "pleadings" (strictly under CPR they are no longer called this but for the sake of convenience I will do so). They will include particulars of claim and a defence and possibly further sequential exchange of particulars (eg counterclaim and defence) and claims further defining the issues between the parties.
- Copies of reports and correspondence relevant to the claim, eg the original valuation report upon which the claimant relies and is now seeking a redress for alleged failures to provide competent advice.
- Copies of relevant legal documents, such as leases, title documents and contracts.
- Relevant copy documents and correspondence from the files of the instructing party.
- If already received, documents obtained on disclosure from the other party.

- If the expert is appointed late in the proceedings, witness statements of fact.
- Copies of expert witness reports already submitted under court directions.
- Other copy documents relevant to the case.

Even under the strictures of limited disclosure of CPR, be aware that the amount of copy documentation in a major case can be extremely voluminous.

One of the first major tasks for the expert is to study the documents, taking special note of those that are particularly relevant to the advice and evidence he is required to produce. There is no short cut here. The expert is expected to have familiarised himself with all the relevant background papers that will be put before the court and the instructing solicitor will expect that the expert will have a good working knowledge of all background aspects. In a recent case, there were 144 ring-binders of documents to be inspected!

It is helpful, at this stage, to prioritise the most important documents in a form of index for future reference during the formative stages of the report preparation.

Finally, it will have been observed in chapter 4 that if appointed within the requirements of CPR, the Civil Justice Council Protocol advises on the procedure to be adopted when seeking to withdraw from an appointment (appendix 2).

Proceeding with the Appointment

The preliminary steps

As highlighted earlier, the first major task is reading all the documents supplied and applying to those instructing you (as usually will be the case) if there is information lacking, or if some of the documents are incomplete. This may take many weeks to resolve since it may involve negotiations between the parties within the general rules for disclosure. It may be necessary for an application to the court to be made for an order for that disclosure.

Importance of case familiarisation

Surveyors new to the litigation field have often questioned the necessity of reading all of the case documents but, from experience, sympathy lies with instructing solicitors since it is difficult to find somewhere to draw the line. The greater the understanding of the background to the case and the events leading up to the claim, the better the expert will be able to fulfil his instructions. It will also assist in the eventual preparation of the evidence and presenting it in the most logical format for the benefit of the court.

It must also be recognised that the expert may also be undertaking an adviser role, particularly in the early stages. This may involve making recommendations within his particular expertise as to how the client's claim or defence objectives are to be met, working closely with the legal advisers. In the more complex cases, it will be extremely difficult to advise adequately without a comprehensive grasp of the case issues, facts and circumstances. Ultimately, the difference between a well-briefed expert under cross-examination and an expert with only a moderate understanding of all the circumstances can, in my experience, make a significant impact. I was once in attendance in the High Court where a well-briefed expert had, very respectfully, responded to the judge that an important question put to him by counsel in cross-examination was based on the wrong facts as evidenced in court documents. This hardly assisted the poor barrister with his further cross-examination endeavours.

Disclosure

Reference has been made to reliance on disclosure of documents that are needed by the expert in order to complete his understanding of circumstances relevant to the production of the expert evidence. It is worth the expert having an understanding of the new rules under CPR and the scope of disclosure that is now available.

The four classes of documents that can potentially be obtained on disclosure in a multi-track case are briefly:

- Those upon which the party relies in support of his case.
- Those that adversely affect the party's own case or support the other party's case.
- Others that are relevant and necessary as part of the "story" or "background".
- Possibly documents that lead to a "trend of enquiry", which might help the party advance his case ("Peruvian Guano" documents).

Leaving aside Peruvian Guano documents, those which are said to be relevant and necessary as part of the "story or background" are clearly going to be the most controversial in practice. Unfortunately, it is often this type of document that will supply important information, which will assist the expert valuer in the preparation of his valuation.

For the purposes of disclosure, a "document" can be paper, computer data or other records, such as tapes, e-mail messages and the like. Each party is under an obligation to disclose unhelpful documents and undertake reasonable searches, but only within the general doctrine of "proportionality". Each party prepares a list of documents that are exchanged together with a disclosure statement, which sets out the extent of the search and the declaration that it has been undertaken to the best of knowledge.

It is still open to the parties to apply for orders for specific disclosure of particular documents or classes of documents or for the carrying out of a search, if standard disclosure is believed to have been insufficient.

The expert will not be involved in the actual disclosure procedures. However, the application to the court may well be as a result of his conviction that special documents or parts of documents are missing that are wholly relevant to the area of expert evidence to be provided. It is my experience that, despite the declarations to be made, there are occasions when disclosure requirements are not fully met, for whatever reason.

The expert's experience and knowledge will provide the grounds for pursuing the specific disclosure of vital documents. However, there will be occasions where incomplete data on a specific issue will mean that the expert cannot properly comply with the *Ikarian Reefer* principles and his duties within CPR Part 35. If so, the expert must discuss fully with those instructing him with a view to pressing for disclosure or, if still not satisfied, a direct application to the court (CPR Part 35.9 and 35.14).

Investigations

The expert and his valuation support team (if appropriate) will be involved in the early stages of making all necessary investigations in areas such as comparable transactional evidence, statutory and planning enquiries and liaison with other experts, eg in relation to building costs.

It is important to observe the rules of confidentiality when pursuing enquires with third parties. It would normally be sufficient to give the preparation of a valuation as being the reason and, if pressed, perhaps in connection with litigation, but that is all. Under no circumstances should the expert or his team pursue enquiries with the other party's legal adviser or experts appointed without prior reference to those instructing him. If it is revealed that evidence relevant to the case is held by the other side or their surveyors, the proper course is to route the enquiry through the instructing solicitor, who will normally be able to obtain disclosure.

In my experience, it is often the case that a valuable piece of transactional evidence will not be disclosed by a third party because of a confidentiality agreement. This is particularly so in the realm of rent review arbitrations. If the court is satisfied that the evidence is relevant material to the case and required in the course of justice, an order for disclosure may be obtained. The mere threat of applying for an order is often sufficient to procure the details.

I have experienced a case where a property agent would not disclose details of a transaction because he was aware of the purpose behind my enquiry, possibly as a result of contact with the other side's surveyor. Through some misconceived display of "loyalty", he would not provide the details. It was sufficient for the instructing solicitor to have a polite word with him and to tactfully mention the powers of disclosure. How-ever, be aware that there is a possible defence to such an application as a result of the decision in *South Tyneside Borough Council* v *Wickes Building Supplies Ltd* [2004] NPC 164, which is reviewed in chapter 16.

Recording evidence

When recording comparable evidence and other facts, be certain to prepare comprehensive file notes giving the time and date and identity of the person supplying the information.

In the case of rental transactional evidence, ensure that the agent or surveyor is asked to complete and sign the RICS pro forma, which is available for this purpose.

Where the evidence provided is substantial, such as in the case of responses from a planning officer at a formal meeting, it may be appropriate to prepare a simple form of witness statement or arrange for the instructing solicitor to adduce the evidence in this way. Much of the factual evidence may be hearsay but, in the case of a civil action, admissible. In the event that such facts, for whatever reason, are not agreed between the experts on both sides, any documents or notes will lend credence to the evidence so adduced. This will assist the degree of relevance and weight that is attached

Inspections

The expert will be inspecting the subject property and those comparables upon which he mainly relies in arriving at the valuation opinion. It is important to inspect the whole of the property and not to omit any parts.

Many years ago, I was involved in a major planning inquiry when another expert suffered the consequences of not adhering to this fairly obvious requirement and, in doing so, caused me considerable (professional) hardship.

The expert concerned was the senior partner of a well-known major firm of quantity surveyors. The planning inquiry was considering an appeal in respect of an application for a listed building consent to demolish or alter a large estate of poor Georgian or Georgian-style buildings in the West End of London.

I was the valuer appointed to give expert evidence in support of the appellant's case that the run-down estate of buildings within a conservation area could only be revitalised economically by a wholesale scheme of refurbishment and redevelopment requiring listed building consents.

My valuation evidence entailed the preparation and production of a large number of detailed development cash flows, incorporating various degrees of refurbishment/adaptation. I was to rely upon the previous building cost evidence to the inspector provided by the quantity surveyor concerned.

The first question put in cross-examination by counsel for the planning authority concerned the degree of inspection undertaken. The questions and responses, as I recall, were roughly as follows:

Counsel: Would you please confirm that in the preparation of your building cost estimates you have inspected internally and externally every part of every building within the estate?

Quantity surveyor: (*Long pause*) Sir (*addressing the inspector*), I have inspected a large number of buildings on the estate representative of the whole but not every part of every building. I left some of this work to a team of assistants within my practice.

Counsel: Did you inspect, for example, numbers 16, 17, 18 and 19 Consort Square?

Quantity surveyor: Sir, I inspected the first two of those internally but viewed the others only externally.

Counsel: (*Shuffling his papers with significant stare at the inspector and a long dramatic pause*). You are expecting this inquiry to rely upon your opinion evidence of building costs in respect of the state

	of these buildings some of which you have not inspected internally?
Quantity surveyor:	(*By now looking rather sheepish. Counsel for the appellant trying to hide his frustration. A long pause*) Well, yes, I believed that an external inspection was sufficient. Of course, members of my team have seen all parts.
Counsel:	But you haven't and you are expecting the inquiry to rely on your team's findings and not your own to that extent?
Quantity surveyor:	Yes (!)

The remainder of the cross-examination can probably be imagined. It was all very painful for the expert thereafter. Every time he was asked about his evidence, the same difficulties of non-inspection arose and, as a result, the credibility of his evidence suffered badly.

As a result of the expert quantity surveyor's lack of proper preparation, my own evidence given over one-and-a-half days necessarily suffered. The costs I was relying upon were often in doubt and (probably quite wrongly) indirectly cast a severe shadow over the reliability of my valuations for that reason. As can be imagined, my pain and frustration as a result of another's neglect has left an indelible impression and an important, if obvious, lesson.

If by chance access to any part of a property is not available to you, perhaps because of security arrangements or a locked door in a basement storeroom where the key could not be found (both cases I have encountered), do supply reasons in your report. Do say to what extent you believe your expert views may have suffered in consequence (if at all).

When making property inspections, do not forget the importance of photographs and, on occasion, illustrative drawings and perhaps, in some major cases, models. Often such exhibits will considerably assist the court in a better understanding of the property valuation issue and your evidence. In an era of digital imaging, it is possible to enhance or alter photo images and it may be advisable to formally agree a set of photographs or other aids with the expert on the other side to avoid any uncertainty as to whether they represent a true likeness of the subjects concerned.

I am often amused, as a rent review arbitrator, by the landlord's expert skill in photographing the subject office building in the golden light of a late afternoon sun, whereas the tenant valuer managed to take a photograph in the gloom of a rainy morning!

Recording time costs

It will be appreciated that the court (or arbitrator) has ultimate discretion over costs. The costs to be paid by one party to the other may therefore become an issue and the expert could be asked to substantiate the amount of his costs. Therefore, it is imperative that an accurate time-recording system is established. This should extend not only to the number of hours expended in any day but also in general terms how the time was used.

The expert is entitled to all his other costs reasonably incurred in accordance with the agreed conditions of engagement, ie the contract between the litigator and expert. Thus, if the court were to ultimately award a reduction in costs, this would not normally hinder those contractual arrangements. However, the client may take a very dim view of excessive charging, which has not been countenanced by the court, especially if this could have been reasonably avoided.

It is strongly recommended, therefore, that on appointment the expert has either a manual or computerised file-recording system and provision for inputs to include a description of the work undertaken on a day-to-day basis. Litigators will often require experts to supply a supplementary schedule showing how the hours were spent at the issue of each invoice.

Fee "budgets" or estimates

In the case of major civil litigation, it is my experience that experts are asked to produce a non-binding estimate or "budget" of costs at the outset. Indeed, the provision of an estimate is recommended in the Civil Justice Council Protocol (see chapter 4 and appendix 2). The purpose of this is twofold. First, the solicitor will generally wish to have an appreciation when advising his client as to how much, globally, the action is likely to cost as part of the overall litigation strategy and for the purposes of assessing the overall risks. Second, CPR Part 43.6 stipulates that the court has the ability to require an estimate to be provided of costs incurred or to be incurred at any stage of the legal action.

The estimate of costs is normally broken down into the various anticipated stages of the appointment and must include the costs of any supporting team undertaking preliminary searches and collation of facts. The budgeted categories will typically include:

- Familiarisation of the case and reading documents.
- Applying for further documentary evidence.
- Inspecting the property or properties and taking details, including comparables.
- Undertaking all local and national market or market-related enquiries.
- Attendance at pre-trial hearings, case management conferences and conferences with counsel.
- Attendance at experts' meetings and drafting statements of agreed and unagreed facts.
- Preparing the expert's report.
- Attendance at a conference with counsel with regard to any clarification.
- Possible further attendance at a meeting of experts and the drafting of any statements of any unagreed facts. Further reports in rebuttal.
- Attendance at mediation.
- Attendance at pre-trial review or meetings prior to a court hearing.
- Attendance at the trial.

It is vital that any estimate or so-called "budget" of costs should be indicative only. It should be emphasised that the costs given are not legally binding and are given for approximate guidance only.

Time costs for litigation are notoriously difficult to estimate. In practice, the case will often develop in ways that were not originally anticipated and the extent of the work will vary accordingly. I have been involved in cases where the final costs were more than double the original estimate! Only in exceptional circumstances be persuaded to agree a "cap" on costs or a specific budget. This may be appropriate in very low-value cases but the limitations of what can be done should be explained.

Meetings and Conferences 9

I briefly referred earlier to the importance of expert meetings and the duty to agree facts and narrow the issues as far as possible. It will be recalled that, under CPR, the court may direct that a meeting of experts takes place in order to identify and discuss the issues. It may specify the issues to be discussed and may direct that experts produce a statement of those issues on which they are agreed and those upon which they are disagreed.

Code of Guidance — **expert meetings**

The Civil Justice Council Protocol (see chapter 4 and appendix 2) goes somewhat further by providing detailed guidance on arranging discussions between experts, identifying and narrowing the issues, reaching agreement and recording points of disagreement. Action should then be taken to resolve the outstanding points. The parties, their lawyers and the experts should co-operate by producing agendas for discussion between experts, which may include questions that enable the experts to state the reasons for their agreement or disagreement.

Interestingly, it is advised that:

> The parties' lawyers may only be present at discussions between the experts if all the parties agree or the court so orders. If lawyers do attend, they should not normally intervene except to answer questions put to them by the experts or to advise about the law.

Since the code also applies to those instructing experts as well as experts themselves, it should be anticipated that this guidance would be complied with. Sadly, in practice, this has not often been the case. Clients should also not attend.

Agreeing facts and narrowing issues

I referred earlier to my particular concerns in cases involving property valuation issues, where expert valuers are not given early enough opportunity to meet and agree facts, narrow the issues and explore areas of disagreement. This is often because those instructing the expert are apprehensive of the effects of exposing their expert to the other side in the early stages of the action. It may be perceived that the case or parts of the case could be prejudiced within the overall litigation strategy.

This is a particularly sensitive area and, in practice, fraught with problems. There is no doubt that, denied such opportunities, the expert may not be acting within the spirit of CPR and his duty to the court. However, many experts will not wish to unduly pressurise those instructing them given the importance of the continuing relationship and possibly, in many cases, the prospect of further instructions in the future.

Of course, if the position becomes extreme, the expert may feel that his position and duty to the court are being severely compromised. There is then probably no alternative but to be resolute and to at least remind the instructing solicitor of the ability of the expert to apply direct to the court for the necessary directions.

Unfortunately, it is my experience that, on occasion, judges themselves are not particularly adroit when it comes to realising that an early meeting of valuation experts can not only assist the course of justice, but also will almost certainly achieve substantial cost savings.

I have often been instructed as expert when the court has directed that the expert should meet after the exchange of experts' reports. Unsurprisingly, the experts' valuation reports are often based to a degree on different facts, assumptions, dates, etc (not only floor areas, but this is the most common example). As a result, the experts have to completely redraft their experts' reports with considerably increased costs and sacrifice of time within the directed programme. The difficulties stem, it is suspected, from pre-CPR attitudes with regard to the use, purpose and loyalties of the expert and a lack of recognition that his primary duty is now to the court and not to those instructing him.

There really should be no problems on this issue if lawyers and others instructing the experts take the necessary steps, on the recommendations of their expert, to prepare an agenda of the items to be discussed, all of course on a without prejudice basis. CPR Part 35.12 (4) provides that the content of a discussion between experts shall not be referred to at trial unless the parties agree. Therefore, discussions are, in any event, wholly privileged. However, CPR Part 35.12 (5) also states that where experts reach an agreement on an issue during their discussions, the agreement shall not bind the parties unless the parties expressly agree. This is a further safeguard.

Another way of resolving the difficulties is to suggest that a formal agenda is agreed between the parties as to those issues that should be discussed by the experts in accordance with the Civil Justice Council Protocol recommendations (see chapter 4 and appendix 2). However, paragraph 18.7 in the Protocol states that "those instructing experts must not instruct experts to avoid reaching agreement (or to defer doing so) on any matter within the expert's competence. Experts are not permitted to accept any such instructions". Thus, it must be borne in mind that any agenda must be subject to that potential departure.

It has been suggested that a further way of avoiding the perceived risks is for the solicitors for each party to attend the meeting of experts. Such an approach is not recommended since, in my experience, it may introduce an atmosphere of rigid formality into what should be a relatively free-ranging opportunity to agree as much as possible. In fact a failure to adhere to this clause may well result in any perceived benefit, as far as the instructing lawyers are concerned, being lost.

In a leading clinical negligence case *Hubbard* v *Lambeth, Southwark and Lewisham HA* [2002] Lloyd's Rep 8 the claimant's lawyers were concerned that at a future experts' meeting their experts would be reluctant to address the actions of the eminent consultant, who was the defendant. They wanted the lawyers to be present. The court upheld the principle of lawyers' non-involvement. They should prepare a "well-crafted agenda".

This will continue to be a controversial area of practice for the expert. If the opportunity to agree facts is not being given and this is perceived to be in breach of the duties under CPR or there is a genuine belief that cost penalties will be involved, it is up to the expert to make this clear to those instructing him. At least, if ignored, he can hardly be blamed later for additional costs being incurred and maybe for being unable to respond within deadlines when further experts' reports have to be produced.

Assuming that the valuer has managed to agree that an experts' meeting is held, hopefully without lawyers, it may be helpful to prior agree with the expert on the other side the extent of the agenda (with the assistance or co-operation of the lawyers as recommended in the Civil Justice Council Protocol), the scope of issues to be discussed in advance so that the maximum benefit can be achieved. I have often attended experts' meetings when, for whatever reason, it has not been possible to obtain the maximum benefit because the other side has been inhibited, either because of the precise instructions given or due to insufficient notice.

Inevitably, in the course of discussions it may become clear why the experts are at variance (if indeed they are). This may lead on to further discussions, which open up an exploration of the issues in conflict.

Possible agreements

Bearing in mind the benefits of privilege and the duty to the court, the expert should not necessarily shy away from the possibilities of reaching agreement. However, it will be appreciated that the expert's duty to the court only extends to an expert opinion and not resolving the subject matter of the action (if different). This should only be a matter for the discretion of the parties and if not agreed, ultimately, the judge.

In practice, it is often found during an experts' meeting that the instructions given differ. Therefore, it is not surprising that agreement cannot be reached since the experts are approaching the issues from a different direction. If nothing else, the experts' meeting will have identified the differences, which may result from misunderstood facts or a different interpretation of the relevant law. Either way, the experts' meeting should be discontinued and only resumed when the parties' conflicting instructions are clarified. I have been an expert in a case where this happened. In fact, the experts were agreed, it was just that they had been given entirely different instructions to follow, one of which was in conflict with facts of the case. The experts' meeting had performed a valuable function leading, ultimately, to the case being settled.

Where it appears to both experts in a civil action that as a result of agreeing a valuation issue this might lead to a settlement between the parties, it is important to explore this possibility (indeed, there is a duty to do so) but under no circumstances should any "settlement of the action" be reached. The experts (as experts) have no such power or duty without instructions.

This will not apply in the case of, for example, a rating appeal or rent review, where the experts will have already been negotiating the case and will have already resolved that they could not agree.

In a civil action, the experts are under a duty to report the possibilities to their respective clients and to seek instructions. If, perchance, some discussions lead on to a possible basis of settlement, it may be advisable, at this stage, that such possibilities are not disclosed in detail since this could violate the "without prejudice" safeguards and possibly undermine the negotiating position of the instructing solicitors. Until clear instructions are given as a result of a discussion between the parties' respective legal advisors, it is best to not take matters further. It is appreciated that this may appear to be in slight contravention of CPR and the Civil Justice Council Protocol but not, it is suggested, if it is genuinely believed that the prospects of settlement might be endangered.

Likewise, notes taken of a discussion between experts are protected under legal privilege. If matters have been discussed beyond the agreement of facts and narrowing the issues, these should not appear in the expert's report. However, matters agreed and unagreed will form the subject matter of a report, but only if jointly agreed by both experts.

Statement of agreed and unagreed issues

The Civil Justice Council Protocol advises that at the conclusion of the experts' discussions a statement of the matters agreed, unagreed (and the basis of these) should be prepared and signed together with any issues arising not in the original agenda plus recommended further action (see chapter 4 and appendix 2).

Modifying the opinion

Having discussed the issues with the expert on the other side, it has not been unknown for an expert to realise that he should have reached a different conclusion. There may have been a misunderstanding of the facts, or new evidence or a different legal interpretation. The expert should not be tempted to soldier on with his original opinions in the face of such a difficulty. If this was as a result of a mistake or error, it is certainly embarrassing, but probably not actionable given that the expert is generally immune from suit. Hopefully, in most instances, this will arise because the new factual evidence has been produced.

The expert's duty is to report this to his client. It is suggested that any advice or reports produced should be amended, which might have the effect of inducing a settlement of the case.

Conferences with counsel

In the course of a major legal action, the expert can be expected to attend a number of conferences with counsel and the instructing solicitor, possibly with other experts in attendance. In the course of such conferences, the expert will be exposed to discussions on case strategy and he may be asked to give advice on some aspects. It is important to be clear as to how to perform the dual roles and to be absolutely clear in responses as to the context in which they are being made. For example, the expert may be asked to advise on the presentation of valuation issues in the best possible light. This should not, in any way, inhibit the expert in emphasising his honest, independent and objective opinions, if it is perceived that the lawyers may be inclined towards other expectations as a result of the advice that is given.

Inevitably, there will be occasions where pressure is brought to bear upon an expert at such a conference in regard to the way in which he might seek to illustrate his opinions in the eventual report or, even worse, alter or colour the evidence. Needless to say, any such suggestions should be stoutly resisted being wholly in breach of the expert's overriding duty to the court (CPR Part 35 (1) and (2)), the *Ikarian Reefer* judgment, the Civil Justice Council Protocol at paragraph 15 (see chapter 4 and appendix 2) and, if relevant, the *RICS Practice Statement*.

Under CPR, such attempts are not usually as blatant as they used to be. Nevertheless, within the context of the litigation "team spirit", it is easy to succumb to even mild suggestions without thinking through the possible consequences. A tactful explanation that any such departure could be seriously prejudicial under skilful cross-examination will usually resolve the difficulty.

There are other ways in which conferences with counsel can be extremely useful and productive to the expert. Knowledge of the case background and experience will be enhanced, leading to greater confidence in the production of a report or giving evidence.

Lawyers can suggest ways and means of producing the evidence in a clear and more concise form and, in my experience, have considerable skills in identifying the important issues and those that can be discarded. The advice received can often lead to considerable time savings and a better appreciation of how to present the evidence required. It may also be necessary to seek advice on legal interpretation, for example in the correct analysis of complex lease covenants. It is rare that I have come away from a conference with counsel without gaining beneficial information, advice or insight that will assist the process of producing a better-quality expert's report.

The case conference will also enable the expert to liaise closely with other experts, whose evidence may relate to, or be in support of, his own. It will be important to ensure where, for example, the valuer is relying on planning or cost evidence for the purposes of arriving at a valuation, that they are working from the same facts, dates and possibly assumptions.

There may be a sequential problem where their evidence is required for the valuer's report or vice versa and therefore deadlines need to be agreed.

Mediation

Even prior to CPR, alternative dispute resolution (ADR), principally in the form of mediation, was becoming more established as a means of inducing the settlement of a dispute without resorting to the relatively costly and time-consuming business of litigation.

It will be recalled that within the CPR "overriding objective", one of the key points is encouraging and facilitating the use of ADR procedures, if appropriate, and helping the parties to settle the whole or part of the case. The court has powers to impose cost sanctions in cases where a party refuses to mediate. In *R (on the application of Cowl) v Plymouth City Council* [2002] 1 WLR 805, Lord Woolf gave judgment for the purposes of a judicial review that:

> both sides must by now be acutely conscious of the contributional alternative dispute resolution, to resolving disputes in a manner which both meets the needs of the parties and the public and saves time, expense and stress.

Encouragement to mediate

In a more recent case, *Dunnett* v *Railtrack plc* [2002] EWCA Civ 303, the successful party in an appeal refused to mediate and was not awarded its costs.

In *Royal Bank of Canada* v *Secretary of State for Defence* [2004] IP&CR 28, Mr Justice Lewison believed that there should be mediation. Although the case was principally about the interpretation of a lease, the defendant won but would not mediate and, as a result, did not recover his costs. Pressure to mediate was also a dominant factor in other reported cases.

The cost sanction

A recent Court of Appeal decision in *Halsey* v *Milton Keynes General NHS Trust* [2004] 4 All ER 920 has cleared up some of the uncertainties. A number of points of clarification, particularly on the issue of the effect on costs awards, can be identified:

*   The court cannot compel the parties to mediate.
*   Litigants must routinely consider with their advisers whether the dispute is suitable for ADR.
*   Some cases may be inherently unsuitable for mediation. Examples include construction, fraud and purely legal interpretations.
*   The onus is on the unsuccessful party to show that the other party acted unreasonably in refusing to join in ADR in order to have a costs award reduced.

The following factors should be considered when deciding whether ADR has been unreasonably refused:

*   Whether the distressful party believed it had a strong case.
*   Whether other methods of settlement were attempted and failed.
*   Whether the cost of mediation would have been high.
*   Whether mediation had been suggested late.
*   Whether the unsuccessful party is able to show that mediation had reasonable prospects for success.

It is made clear that if a party does not attempt ADR in contravention of a court order it may be regarded as unreasonable. The result of this decision is that the courts will not generally compel the parties to attend an ADR. In many cases, it is an important alternative and may have cost implications where a successful party did not attempt it.

It follows from the foregoing restatement of the relevant law that mediation will, in many cases, be a part of the litigation process. The expert can be expected to be involved as an adviser in such proceedings (although he may not always attend).

Nature of mediation

As has been stated earlier, the purpose and exercise of mediation is entirely different from any form of litigation. Mediation is not about a third party making judgments about the case but attempting to draw the parties close together so that they can be more easily placed in the position of reaching a settlement.

Therefore, the mediator is in the business of breaking down the barriers, throwing fresh light on the apparent differences, and, very often by emphasising the objective weaknesses of a party's case rather than its strengths.

I have attended a number of mediations and the procedures are usually more or less the same. Generally, the mediator starts by setting out the procedures and objectives, reviewing his understanding of the case and the issues in dispute. He will invite the parties to summarise their respective cases but not in a strictly evidential sense. He will then "shuttle" between the parties, concentrating on specific issues and trying, piece by piece, to find common ground.

At times, the mediator can be very robust in the way that this process takes place and, conversely, he will sometimes exercise extreme diplomacy. The whole procedure is entirely private and without prejudice. Part of the mediator's task will be to build a sense of confidence that a settlement may be achievable as a result of confidential discussions initially through him as an intermediary. Where these discussions appear to be positive and potentially productive in terms of finding common ground, he may invite the parties around the table under his chairmanship.

The expert's role

The expert as adviser will inevitably play a key role both in the interviews with the mediator and possibly also with the other side, as the process develops. It is important, if taking an active role, that the expert "adviser" understands fully his instructions and the degree to which the spirit of co-operation and openness leading to the potential for settlement is to be embraced.

To have any chance of success, the parties and their advisers will need to be positive in their desire to settle and their approach to the whole process in a positive and co-operative fashion. The emphasis is on a non-confrontational approach, seeking the parties' resolve to look beyond their differences in a problem-solving sense.

Normally, the expert's adviser will be instructed or briefed on how the party sees his role and the extent to which he will be asked to enter into the process. If not, the expert should certainly be seeking such clarification. It is impossible to negotiate by committee and therefore important to clarify whether the expert is to lead the discussions on some issues or merely act as an adviser to the lawyer or whoever is representing the party as the process continues.

In many cases, mediation is successful. It is believed that a significant proportion of civil actions are settled out of court as a result of attempts to mediate. According to statistics provided by the Centre for Dispute Resolution (CEDR), over 90% of mediations are successful.

If an agreement results, this will nominally be drafted into a legally binding agreement setting out the terms of settlement.

Most value-related disputes are suitable for mediation but some more than others. Multi-issue disputes with more than two parties are particularly suitable but not cases where legal precedents are required.

The Expert's Report

It is likely that a valuation surveyor who is instructed to provide an expert witness report will already have considerable experience of report writing. However, there is a need to adopt a different approach. For example, there are specific requirements in regard to order, compliance, layout, declarations and presentation. Subject to having certain key ingredients, the drafting of a report is still largely a matter of individual style and suggestions can only be made in this regard.

Compliance

Paragraph 13 on page 22 of the *RICS Practice Statement and Guidance Note* helpfully summarises those matters that are mandatory, as far as content is concerned, and is a useful reminder. The *RICS Guidance Note* goes on to provide an example of structure and nature of content but comments that variations to the structure are required according to circumstance. The Civil Justice Council Protocol, referred to in chapter 4, expands upon the CPR requirements (see appendix 2 for the full text). In brief:

- The content and extent of an expert's report is governed by the scope of instructions, general obligations (including the overriding duty to the court) and CPR 35 and PD 35.
- Experts to maintain objectivity and impartiality.
- Address to the court.
- "Model" reports available from the Academy of Experts and Expert Witness Institute (NB: there is no requirement to use these).

- Reports must contain statements that they understand their duty to the court and have complied with that duty. The form of mandatory statement of truth (which cannot be modified) is as follows:

  I confirm that insofar as the facts stated in my report are within my own knowledge I have made clear which they are and I believe them to be true, and that the opinions I have expressed represent my true and complete professional opinion.

- Expert's qualifications to be included must be commensurate with nature and complexity.
- Include qualifications of others who have provided opinions in support.
- Keep fact and opinion discrete and separate.
- State facts upon which opinions are based. Distinguish between facts known to be true and those assumed.
- Where facts are in dispute, express separate opinions on each hypothesis.
- Where there is mandatory range of opinion based on published sources (eg property market indices), state the qualifications of originator, particularly if representing a well-established school of thought.
- Where there is no source for the range of opinion, the expert may express opinions on what is believed to be the range other experts would arrive at if asked.
- Must include a summary of conclusions at the end of the report and possibly also in an introductory "executive summary" if there are complex matters (see later in this chapter).
- The mandatory statement of substance of all material instructions should not be incomplete or otherwise tend to mislead.

Leaving aside the mandatory compliance matters as set out above and reviewed earlier in this book, it will be of assistance to give some general guidance on how to approach the drafting of the report given the audience to whom it is addressed (particularly the court), its purpose and use.

The expert witness report is fundamentally different in one important respect from other professional reports prepared by the valuer. It is addressed to the court or tribunal/arbitrator. It is a personal stand-alone document giving an expert opinion that needs to be credible and as robust as possible, given that it may be tested under cross-examination.

It needs to be carefully tailored according to the likely recipient, thus fully explaining technical matters to a county court or High Court judge, with perhaps less explanation to a surveyor arbitrator or professional tribunal.

Legal advice and corrections

In the days prior to CPR, it was customary for instructing lawyers (and in particular counsel) to become deeply involved in the preparation of their expert's report, with many drafts being prepared and corrected. While this did not usually extend to attempts to vary the expert's advice or opinion, it was my experience that sometimes suggestions were made that would give the report a particular "spin" by means of subtle colouring, exaggeration, omission or certain emphasis. Within paragraph 15, the Civil Justice Council Protocol provides very clear directions on attempts to amend an expert's report. In brief:

- Experts should not be asked to, and should not, amend, expand or alter any parts of reports in a manner that distorts their true opinion.
- Experts may be invited to amend or expand reports to ensure accuracy, internal consistency, completeness and relevance to the issues and clarity (see chapter 4 and full text in appendix 2).

It is entirely appropriate for instructing solicitors or counsel to make suggestions as in the last point above. A report may not deal succinctly with the issues and may not be sufficiently clear for the judge and the court to understand the expert's reasoning and opinions. Experienced counsel, in particular, are past masters at identifying the essential matters to be included in a report dealing only with the issues upon which evidence is sought. They will often add considerably to the honing of a report in terms of logic, clarity and brevity. However, under no circumstances be persuaded to vary or amend the essence and scope of the opinions given.

I have known situations where an instructing lawyer has not fully understood the reasons for the evidence being prepared or being supported by certain arguments or technical issues. This is an opportunity to impress instructing lawyers with the expert's knowledge of the particular issues and how they relate to the expert opinions being expressed. Such a dialogue will not only assist in any final amendments but also will inform the lawyer more deeply in his understanding of the expert evidence issues.

Preparation

Assuming that the expert has accumulated all the facts and evidence required and has prepared the valuations, he is then left with the task of preparing the report.

I suggest that the first consideration is the likely state of knowledge of the recipient. This will determine the extent of the report and the explanations given, but also extend to the background information that will precede it. Thus, a county court or High Court judge will perhaps need a fuller explanation of the technical details. This may be superfluous in the case of arbitrators or tribunals.

Do seek your instructing lawyer's advice as to the extent to which you need to include a background case history to the dispute linking into the need for the expert evidence. In a major civil action, either a skeletal or fully fledged history of all the circumstances will probably have already been given by counsel for the claimant in his opening address. It may only be necessary to briefly refer to this in order to put your evidence in context. Whatever the extent of case history and background, do "tell a story" creating sufficient interest to make the document eminently readable.

Structure

I believe it is helpful to prepare comprehensive headings and notes at this stage. This includes the order in which the facts are to be presented followed by the data supporting your evidence, the evidence itself, your opinions and finally your conclusions. Do reject any matters that are wholly superfluous to the general thrust of the evidence being given. Keep striving for brevity and clarity.

If possible, avoid qualified opinion statements. Remember that the advice and evidence needs to be robust and certain. In some professional reports to a client, the valuer may qualify, to some degree, statements of advice or opinion possibly because of legal liability. An expert witness has immunity from possible claims (see chapter 14). In any event, if any material requires serious qualification, it should not form part of the evidence.

Hearsay

Be mindful of the important differences between proven or agreed fact and hearsay. In civil actions, hearsay will normally be admissible by reason of the Civil Evidence Act 1995 (see chapter 5) or as a result of arbitral or tribunal directions. But bear in mind that the degree of weight to be attached will often depend, to some degree, upon how it is presented in evidence. If you have had first-hand knowledge, for example, of a comparable transaction, this will hopefully have been agreed in advance as a matter of fact, therefore it cannot be challenged. However, if not, then the report needs to indicate its source and extent to which it has been verified (if any). In practice, it is to be hoped that any property transaction evidence will have been agreed between the experts in advance in accordance with the CPR guidance or directions.

Advocacy

Earlier, I reviewed the important differences between expert evidence and advocacy. Do ensure that there is no attempt to be persuasive on any element of the client's case. Only incorporate direct, objective, honest opinions on the issues where expert evidence is required. In addition, you should not indicate or suggest any particularly basis for a judgment, award or determination as this could usurp the role of the judicial body to whom you are reporting.

Technical jargon

Do explain all technical expressions. Where not addressing a report to a fellow surveyor, avoid property industry jargon or vague catchphrases, such as the property is "tired". Always include your area measurements in imperial (sq ft) and metric. Do not try to over impress by including verbose text concerning expertise or experience that are not wholly relevant to the logical development of your evidence. Do not, for example in the case of a negligence action, state what you as the expert would have done. It is only the evidence of opinion and the facts that are required.

I have included a checklist of what to include and what to avoid at the end of the chapter.

Suggested report sections

These are generally in compliance with the *RICS Guidance Note*, CPR requirements and the Civil Justice Council Protocol.

Title page

This should include always the court or tribunal reference number, to whom the report is addressed, the parties, the expert's name (with qualifications), the nature of the evidence being given, the date and on whose behalf it is being given. Appendix 4 includes a a typical title page and headings in the case of a civil action report.

Contents

From my experience, there will usually be two contents pages. The first one refers to the main body of the report while the second details the exhibits or appendices with page numbers and possibly paragraph numbers. Do ensure that the pages are numbered and have paragraph references.

Executive summary

It will always be helpful to include a synopsis of the report detailing, briefly, name and qualifications, the context of the evidence, the instructions given, description of the property and other important details (eg lease, heads of terms), the opinions expressed with brief reason and conclusions.

This will enable the reader to have an "at a glance" grasp of the opinion evidence without the need, perhaps at a preliminary stage, to study the contents in detail.

Experience and qualifications

This should not include a full CV. What is required here is a statement with qualifications, years of experience, specialist expertise and position held. Expand briefly those areas of expertise that particularly relate to the property being valued and the issues.

The expert may wish to include a brief "CV" covering the full details of his career as a valuer in an appendix. The court or tribunal will thereby have a comprehensive grasp of who he is, the experience undergone, how the expert is qualified to give evidence and any specialist expertise that is relevant.

The instructions

As indicated in chapter 4, it is unnecessary to reproduce the whole of the letter of instruction unless the court has ordered it to be produced, only the "substance". However, it will be necessary to include the precise instructions given as to the specific valuation evidence required in your report and any other matters of opinion that have been requested. CPR Part 35.10 (3) stipulates the requirements (see chapter 4 and appendix 1).

The question of the extent of "the substance" and the impact on the wider issue of litigation privilege (see chapter 15 for a fuller explanation of the doctrine) was considered by the Court of Appeal in *Lucas* v *Barking Havering & Redbridge Hospital NHS Trust* [2004] 1 WLR 220.

The court concluded that the intention of CPR Part 35.10 (4) was to encourage the expert to set out all material instructions but defined "instructions" as including all information supplied to the expert that could be considered as being "material". It may not encompass all the information provided.

There are obvious dangers here for the instructing solicitor who inadvertently supplies information to the expert that is material to the instructions but which it had been hoped would be "privileged" from disclosure. If in doubt, the expert should fully discuss the position at the earliest opportunity with those instructing.

Investigations and enquiries

This is an opportunity to confirm the documents seen, the enquiries undertaken and the inspections made. The content here will vary considerably between one case and another depending upon the circumstances and the precise extent of the role.

In a civil action, the expert will usually have been supplied with a considerable volume of case papers. It is unnecessary to refer to all of these but may be sufficient to refer to classes of documents that have been seen. However, always include reference to those documents that are directly relevant to the evidence, for example leases, title documents, planning permissions and other expert witness statements that have been relied on.

I suggest that legal guidance may be needed as to the extent of this disclosure in the case of a major civil action. Refer to the specific documents central to the action, for example the valuation report or reports in respect of which negligence is alleged, contract documents and correspondence in the case of a breach of contract action.

The extent of investigations and enquiries undertaken should be briefly recorded, including the inspections, when these took place and also reference to inspection of comparables as appropriate.

Background history (chronology of events)

It is a CPR requirement to include a "chronology of the relevant events". The extent of this will depend upon the extent of the history given by the counsel in a civil action. However, it is always important to set down the main events — "to tell the story" as it links in with the expert evidence you are instructed to give. This may include a section with reference to the persons involved in the events on a purely factual basis.

Establishing the facts

Again, the extent of the report content under this heading will depend upon the particular case, the issues and the circumstances. In almost all cases, it is an opportunity to record (or refer to) the agreed facts of the subject property central to the dispute. The subheadings might include:

- situation and location
- description (including specification, services)
- repair and condition
- floor areas
- site area(s)
- rating assessment(s)
- lease details (heads of terms)
- rent review clause
- planning.

Where the information to be relied on within any of the above headings is substantial, it is suggested that only a synopsis is included here, with a more comprehensive version in an appendix.

Other relevant facts

In many disputes, much of the above material will have been agreed in a separate statement of agreed facts. Nevertheless, even having perhaps included the statement of facts as an appendix to the report, it is still useful to summarise the main areas of facts so that the report is comprehensive and relates logically to the main body of the text.

Market background and evidence

This section may be subdivided according to need. Its precise structure will again depend upon the nature of the property being valued. In order to support the valuation opinions being expressed, usually it will be necessary to provide a background to the relevant market or markets indicating trends and perhaps referring to indices. However, do bear in mind that indices should rarely provide primary evidence in support of valuation conclusions. Only refer to them in a supportive role. Any supporting, independently published reports of indices and market trends do not need to be proved (if published and in the public domain). However, it is suggested that they are included in an appendix.

Do think carefully about the inclusion of independently published research material on markets. In my experience, they are often too general in their conclusions to be of substantive help in guiding a court or tribunal. As an arbitrator, I have often discovered, after careful and thorough reading of such reports, that they are capable of suggesting the very opposite of the trends being relied upon by the expert!

A substantial trawl of rental or capital evidence transactions may have been carried out. It is unnecessary to include, either in the body of the report or elsewhere, a complete list of these. Try and be economical by selecting the best evidence to which the greatest weight is attached and provide a commentary on each of these. If it is necessary to include a longer list of transactions, it is suggested that these could be included in an appendix but only referred to briefly.

Do ensure that, in relation to the preferred evidence of transactions, sufficient explanation is given as to why the greatest evidence is attached to these and the factors that influenced the selection.

It will usually be necessary to include copies of the floor plans, location maps, ordnance survey extracts, photos and other such documents in the report, perhaps summarised in this section or placed in an appendix. Do remember to obtain all necessary authorities' consents for publication in a report. Given the possibility of digital distortion, agree a schedule of photographs with the expert on the other side.

Expert valuation opinion

This section should develop the detailed valuation "trail" based on the evidence of transactions leading to your valuation(s) and therefore expert opinion. Working as an arbitrator, I have often found that expert valuers fail to sufficiently connect their analysis of the transactional evidence adduced to the adoption of the valuation required in a capital value or rent review context. It is often difficult to understand the process by which the valuer has connected evidence of transactions relied on to the subject-matter being valued. The links between the evidence analysed and the valuation is unclear or missing. If it is important for a rent review arbitrator, it is absolutely essential in the case of a High Court judge.

Many years ago, I was involved in a major negligence case where a High Court judge listened to the evidence of the three expert valuer witnesses over a period of five days on a substantial and complex industrial property investment with many valuation calculations. At the end of the five days, he asked of the claimant's counsel whether he could explain again (since the point still seemed to have alluded him) as to the meaning of "year's purchase". This was truly a "showstopper". The expert witnesses and legal representatives were aghast at this revelation. It came as no surprise that the High Court judgment was riddled with errors and misconceptions. It was successfully appealed, but at great cost and with considerable delay.

Therefore, I cannot emphasise too greatly the need to provide the fullest explanation of valuations, technical terms and the calculations themselves in either the High Court or the county court, although it is suggested that there is less requirement in the case of an arbitrator or the Lands Tribunal.

Most investment and development valuations of commercial property interests are now undertaken using proprietary computerised valuation software. These generate calculations in spreadsheet form that, unfortunately, do not always provide the logical steps in the calculations in a form that can be readily understood by the layman. As such, the expert witness should either adopt a manual approach or provide very comprehensive notes to explain how it works.

Other opinions

In some civil actions, the expert will have been asked to comment in instructions on other issues that are within the area of expertise but may not be directly related to valuation points. For example, whether the other party could have undertaken a scheme of development in the market circumstances at the time given the poor letting potential. The valuer should not be tempted to stray too far into areas caused by other surveying disciplines, for example letting agents, unless he is wholly satisfied that the relevant knowledge and experience truly qualifies the valuer to be regarded as an expert. If limitations or qualifications are to be expressed, it is better to decline giving this opinion.

In this part of the report and later in the conclusions, the valuer must remember, that where a range of opinion is possible, to summarise the range and give reasons for selecting the opinion. This will usually be the case where property valuation evidence is being given.

Markets vary in their degree of volatility, as evidenced by transactions, and there will usually be evidence of variations. Within those variations it is usually possible to come to one or more differing opinions. The courts have referred to this phenomenon (in my view quite wrongly) as the "permissible margin of error". Given the unfortunate inference of "error", it is perhaps unfortunate that they have not referred only to "admissible market variations". As discussed earlier, enhanced credibility can be earned by a solid analysis of the transactional evidence variations with, hopefully, logical and robust reasons for the degree of weight that is attached to the chosen evidence.

Conclusions

In this section, the expert is reviewing the valuation and there may be other opinions reached, the reasons for such adoption and stating the conclusions that can be drawn from them. It may be appropriate to include here the alternative opinions that could be reached and the reasons why a particular basis of valuation is preferred. The expert should not be tempted to usurp the court or the arbitrator's authority and role by including any judgmental conclusions. For example, stating, in the light of the valuation opinion reached, that the defendant was negligent. In the case of a rating appeal or rent review arbitration reference, the valuer may wish to urge the arbitrator or tribunal to reach a particular determination or award, but this should not be included in expert testimony but as part of a separate case statement (presented as an advocate).

Declaration and compliance

Expert witnesses differ in their positioning of the declaration and certificates of compliance. Sometimes it is included at the front of the report, sometimes at the end. The latter is generally to be preferred, but note that the statement required in CPR Part 35.10 and in the *RICS Practice Statement* must be at the end.

Here it is necessary to include the declarations referred to earlier under CPR, including the statement of truth, declarations required in the *RICS Practice Statement* and, in particular, include CPR Parts 35.3, 35.10 (10) and 35.10 (2). The whole of Part 35 is reproduced in appendix 1. A suitable form of declaration covering all compliance requirements is in appendix 4.

Checklist

In order to guide the valuer expert in the preparation of a report, set out below is a checklist containing some of the essential features and elements, the aspects to avoid and the matters normally to be included according to the circumstances.

Table 11.1 Report checklist

Make your report:	Avoid:
Personal	Hearsay
Honest	Advocacy
Objective	Misrepresentations
Clear	Exaggeration
Unambiguous	Unrealistic assumptions
Robust	Jargon
Comprehensive	Repetition
(as to fact and opinion)	Dishonesty
Concise	Complex technical issues
Credible	Hypothesis
Logical	Judgments (but weight the issues)
CPR/RICS/code-compliant	Ambiguity
Consistent	Inaccuracy and inconsistency

Table 11.2 What to include in the report

Title page	Address to the court
Fact before opinion	Executive summary
Documents seen	Inspections undertaken
Agreed facts	Qualifications and experience
Chronological case history	Definitions and source
Description and facts of property (brief)	Advice and evidence relied on
Sufficient technical explanations	A property valuation "trail"
Full valuation calculations	Realistic assumptions
(possibly in appendices)	
Hearsay with supporting documents/evidence signed pro formas	
Commentary on possible alternative opinions	
Declarations and certificate —	
Protocol/CPR/RICS etc	
Supporting documents (details in appendices)	

I have included a short example report (headings and compliance paragraphs only) in appendix 4. All case references are wholly fictional.

Reports in rebuttal

Following the exchange of experts' reports, the parties will wish to have comments on any differences of fact or opinion. The court will usually direct that a further set of reports is prepared with an explanation of the matters agreed and unagreed in the hope that some issues may be narrowed.

In my experience, some surveyor experts wholly misinterpret this requirement. This is not an opportunity to attack the views of the other expert with criticisms, scoring points and making cross-examination arguments. The primary duty of the expert is still to provide honest, objective opinion, assisting the court with the apparent differences and the reasons for these. Include any conclusions as to why it is believed the other expert has erred in fact or opinion, or possibly revising the opinion if this is deemed to be appropriate.

Where the expert is also the advocate, as will often be the case in a rent review arbitration, the written counter-submission should make it clear when the views of the other valuer are being critically examined in an adversarial manner and when an objective opinion is being tendered as to the reasons for the differences in valuation.

Questions to experts

Within the provisions of CPR 35.6, there is a procedure for written questions to be put to an expert seeking clarification of opinions or issues in his report.

It should be noted that an expert has a duty to provide answers where questions are properly put. Failure to do so might cause sanctions to be imposed against the instructing party. Answers should be copied to those instructing the expert and they are covered by the statement of truth being part of the overall evidence. In practice, such questions would be discussed with those instructing, particularly (as so often) if they are "not properly directed".

The Trial or Hearing

The valuation expert should always assume, when preparing his written expert testimony, that he may be required to present his evidence orally. While a trial or hearing is an uncommon event, particularly in the realm of civil action, post-CPR, nevertheless efforts to settle occasionally fail. The expert will then be faced with the prospect of presenting his evidence and undergoing the ultimate test of credibility by cross-examination.

Cross-examination by an experienced barrister before a judge is likely to be far more challenging than before an arbitrator or tribunal. However, a solicitor or indeed some surveyors can, in my experience, achieve quite exacting performances within the role. Thus, my comments in this chapter apply to all forms of oral hearing and the degree of preparation referred to should be the same in all cases.

Where the expert surveyor is also being relied upon to plead the case, acting as an advocate, this can make the role more complicated and may call for even greater skills in terms of resolving conflicts and a seamless approach. In my view, such a dual role should be avoided. However, this chapter is concerned only with oral expert evidence presentation and not cross-examination.

For many surveyor experts, the prospect of cross-examination before a judge seems to generate a high degree of anxiety. Certainly, it can be challenging but it should not stimulate any irrational fears, as long as the expert is experienced in his field, confident, truthful and, above all, well-prepared. Having the correct experience is fundamental to the role as is the need to be entirely objective, fulfilling the obligations to comply with the overriding objective of assisting the judicial body.

Success will be achieved as a result of experience coupled with thorough preparation and recognition of how to perform in the witness box.

Preparation

Thorough preparation and a grasp of all relevant issues is fundamental to a confident performance. In particular, it is recommended that prior to a hearing the expert should:

- Revisit the instructions (possibly as varied), case notes, transactional and other supporting evidence, statements of agreed facts, the valuations and the report.
- Read again the expert's report submitted on behalf of the other side and any other relevant experts' reports and witness statements. Revisit the most important of all documents that are relevant to the case and that are in the bundle of court documents to which the valuer may be referred. Reinspect and, if necessary, memorise important guidance notes or other background material.
- Liaise with instructing lawyers and, in particular, the solicitor or barrister who will be leading. Find out the names of the legal team and witnesses on the other side. Review the strengths and weaknesses in the expert's report.
- The expert should place himself in the shoes of the cross-examiner. What questions are likely to be put? Rehearse the responses that might be made.

Arrangements on the day

These are points to remember on the day when you are destined to give evidence:

- The expert should liaise with the instructing solicitor/counsel in regard to the start time and the order of evidence to be presented. He should have a contact telephone number if his presence is required earlier or later.
- Ensure that any plans, photographs or other visual exhibits can be made available.
- Arrive in court in plenty of time, preferably during a prior period of statements being made or evidence being given so that you know how the proceedings will go.

- Rehearse the introduction and initial statement. Arrange for the correct type of oath or notice or possible affirmation to be given to the court by the instructing solicitor/usher.
- Be present during the opponent's evidence.
- Remember to turn off your mobile phone!

Personal appearance

The expert will wish to appear honest, open, reliable, professional and confident. Therefore, dress accordingly.

Remember that the judge, tribunal or arbitrator will inevitably assess the expert to an extent on first impressions. It is human nature.

Presentation

When called to the witness stand in court, the expert should proceed purposefully and not too hurried. When arriving, face directly to the judge. The expert will then be asked whether he wishes to swear or affirm or adopt some equivalent practice relevant to his religion? If on oath, the expert takes the Bible in the right hand and reads the following words, which will be on a printed card:

> I swear by Almighty God that the evidence I shall give shall be the truth, the whole truth and nothing but the truth

Alternatively, an affirmation will read:

> I do solemnly sincerely and truly declare and affirm that the evidence I shall give shall be the truth, the whole truth and nothing but the truth.

Having taken the oath or affirmation, the expert could be liable for perjury if any untruths are made in the course of the evidence.

The next step is for the expert to give, when asked, his name and professional address together with qualifications and experience.

In some tribunals and arbitrations the expert may be expected to recite the whole of the written evidence. However, this is rare. Normally the judge, arbitrator or tribunal chairman will have read the evidence and it will only be necessary for the expert to be cross-examined upon it, having first been asked whether he wishes to amend anything.

Oral evidence

First, comments are made upon the general techniques, which may be equally applicable to other forms of presentation, not just giving evidence in court.

It is a good idea to make a conscious point of slowing down the oral delivery. Most people deliver too fast. For presentational purposes, to a court or audience it is better to be too slow rather than too fast. The expert should try to project the voice, particularly if he is normally soft spoken. Do vary the pitch and resonance. It is advisable for the expert to practice this if not normally a public speaker.

The object is to enunciate the evidence or responses in a firm and clear authoritative voice, with as much gravitas as seems natural. Do avoid monotones or constant levels of speech. This can be achieved, to some degree, by altering tone, pitch volume, the speed of delivery and the pauses in between.

If in a court, the expert will have entered the witness box facing the judge. This stance should be maintained whether standing or sitting throughout the session. It is the judge tribunal or arbitrator who will wish to receive the responses, not the cross-examiner. Therefore, glance sideways to receive a question being put to you in cross-examination, but the stance should be maintained direct to the judge or tribunal making as much eye contact as the expert can without being oppressive in this regard. By this means, if nothing else, the expert avoids the cross-examining lawyer's real or contrived expressions of exasperation disbelief or whatever!

Carefully watch the judge or tribunal chair as he makes notes. Ensure that the timing of responses coincides with his own speed so that he does not miss anything.

How to deal with cross-examination techniques

In our adversarial system, the object of cross-examination is to test the evidence. The other side will therefore seek to discredit the expert's evidence, or, at the very least, try to raise serious doubts over credibility. It is important to understand and, therefore, anticipate what is going to occur and the reasons for so doing. It is not a personal assault on the expert or his professional standing, although it is easy to overreact. Bear in mind that this is the only opportunity that the other side will have to undermine the opinions given as an expert witness and also one of the opportunities to enhance their own case.

An experienced barrister may use certain "tricks" and techniques at his disposal, while remaining within court rules and the rules of evidence, to undermine the expert's credibility and competence. Therefore, the questioning can be somewhat aggressive although, in my experience, less so since the advent of CPR. There is a greater emphasis on assisting the court and testing the evidence rather than the arbitrary demolition of an expert, particularly if it means he is not given a reasonable hearing.

Most of the questions asked will be aimed at discrediting the strength of opinion by seeking out the weak points, inconsistencies, potential bias or opportunities for differing opinion. Questions will commonly include:

- Efforts to challenge the degree of expertise, qualifications and ability to provide the opinions expressed, often making unfavourable comparisons with the expert on the other side.
- Questioning independence and integrity.
- Asking questions sometimes in an ambiguous sense in order to create confusion and therefore making you stumble over the true import of the evidence.
- Undertaking a line of examination on issues that appear to be irrelevant. Do not be lulled into a false sense of security since the line of questioning may have an ultimately serious intent. Answer the questions methodically and responsibly.
- Asking questions on matters strictly outside the valuer's expertise. Do not attempt to respond. Explain to the judge precisely the limits on competence, which means that it is not possible to assist the court on those matters.
- Asking a line of multiple questions, again often to cause uncertainty and confusion. Take each question methodically and carefully. If the expert needs them to be repeated, do this through the judge.
- Attempts to rebut the expert's views and methods of preparing the evidence.
- Suggesting there are possible areas for differing expert opinion. In some instances, particularly in regard to valuation evidence, it may be necessary to acknowledge that this may be something to consider. The expert should explain why it is that a particular opinion has been reached and why it is to be preferred.

- The expert may be asked to accept the differing opinions expressed by the expert on the other side. It may be necessary to respond that the other opinion is understood, which may or may not have been validly produced but, for the reasons given, he is firmly of the opinion that the evidence given is more correct.

- An often-practised diversionary technique is to introduce the hypothetical question. The purpose is to draw the expert away from the foundations upon which the evidence is built. Do not reject out of hand the basis of the hypothesis since this may appear unhelpful. If the hypothesis cannot be answered, explain to the judge why. If it can be answered, follow up the line of argument but be careful to explain that the hypothesis must be distinguished from the evidence and conclusions that have been deduced and why.

- Barristers sometimes try to give the impression, in the course of their questioning, that they have a greater appreciation of the technical area (in this case of property valuation) than the expert has. Do not be fooled by this approach. The purpose is a challenge of superiority, seeking to diminish your confidence. Be certain that in responses the degree of expertise is not diminished as far as the judge is concerned.

- Counsel will often try repetitive questions possibly with a view to undermining the expert's patience or demeanour but also, more importantly, to test whether the same responses are received. The cross-examiner will exploit any inconsistencies.

- Experienced counsel, particularly those taking the superior approach to questioning, will seek to undermine the expert's confidence by conjured up signs of exasperation or frustration and particularly by using long pauses. When such a pause occurs after a response, it is tempting to add to the reply. Silence can be unnerving and may lead the expert to the false conclusion that he has not answered adequately. Do not fall into this trap.

The expert should not overreact to any questions, however outrageous they might at first appear. Respond truthfully, after giving careful due consideration, to the judge making whatever explanations are necessary in order to assist the court. The expert must answer questions with reasonable brevity but, as expert, there is a duty to expand if it is believed that an explanation is necessary for the court to fully understand the issue. Do not be dissuaded if, on such an occasion, counsel seeks to terminate your response. This is for the judge to decide.

The expert should not feel intimidated by aggressive questioning. Simply give a well-considered honest and clear response. If a cross-examination question is a fair one, acknowledge it and explain the relevance and effect on the opinions given to the judge.

On occasion, counsel will ask you to comment upon particular documents. If the expert is at all uncertain as to the issues he is referring to, he should ask, through the judge, to be taken to that part of the document bundle that contains the relevant document or paper. Take time, revisit it carefully and respond after due deliberation.

The expert should treat every question as an opportunity to expand further the evidence of expertise and not as a criticism, however negative it may appear. Be positive and authoritative.

The expert may be asked to comment upon the expertise and credibility of the expert on the other side. He should not be led into the trap of personally criticising the professional counterpart. The expert should acknowledge, if it is appropriate, that he has experience in the field and may be qualified to give an opinion. However, the expert should explain to the judge why it is that the alternative basis upon which his evidence has been produced is not accepted and the reasons why the expert opinion is to be preferred. Any attack on personal credibility is a matter for counsel (possibly on the expert's advice).

In cases where your cross-examination has not been particularly productive, counsel may be tempted to launch at you a final desperate question. He will put it to the expert that he is a less than credible expert witness, whose indifferent expertise has produced a wholly unreliable opinion to which little weight should be attached, or words to that effect. It is a form of intimidation but under no circumstances react accordingly. As in all responses, the expert must patiently and carefully explain to the judge why it is that the valuation evidence given has been accurately and correctly deduced as described in the report, resulting in an authoritative and reliable valuation opinion, which it is believed is wholly conclusive.

The expert must not comment upon the weight to which the court should attach to the evidence. This could be interpreted as an attempt to usurp the role of the judge. Finally and most emphatically, he should never enter into any competitive dialogue with the cross-examiner or, for example, anticipate a line of questioning by including a response to the supposed next question. At best, such practices will appear defensive and will tend to harm credibility. In an extreme situation, the court may draw a wholly unfavourable inference.

How to address

County court	Sir/Madam or
	Your Honour (judge)
High Court	My Lord/My Lady
House of Lords/Court of Appeal	My Lord/My Lady
Tribunals	Sir/Madam Chairman
Arbitrators	Sir/Madam
Barristers	Do not address, route questions
	(eg for clarification)
	via the judge or tribunal

The Single Joint Expert

I have referred earlier to the power of the court within CPR Part 35.7 to direct that evidence is given by a single joint expert (SJE). He may be appointed by the court if the parties cannot agree.

In his original reports prior to implementation of CPR, Lord Woolf favoured the wide use of SJEs but his proposals, in this regard, were strongly criticised. Such appointments now only commonly occur in the smaller actions and not generally in multi-track cases.

While there is no doubt that the SJE is able to identify more closely with his duty to the court, there are some distinct drawbacks and potential difficulties.

The courts seem reluctant to impose an SJE on the parties. Perhaps this is an acknowledgement that the procedures required are involved and time-consuming to the extent of almost negating any cost advantages. It is also possible, in the larger actions, that even having joint appointed a single expert, the parties nevertheless appoint their own experts to monitor progress and "audit" the SJE's opinions.

Indeed, in *Layland* v *Fairview Homes plc* [2003] CPLR 19, it was held that the circuit judge was not wrong to reach his judgment as to the diminution in value of a house based on the evidence of an SJE's report. However, since the claimant had obtained different evidence from two other valuers, they should be entitled to exploit criticisms of the SJE by cross-examination. It was for the court to decide the valuation issue, not an expert.

Independence

The difficulties of an SJE appointment arise from the nature of the joint appointment and the independent relationship, which has to be maintained with carefully balanced demonstrable equity to both sides. In practice, it is found that such an appointment can be likened to the skills and subtleties required when appointed as an independent expert (for example, in a rent review dispute), which will be referred to in chapter 16.

It is necessary to maintain an even-handed approach, being seen to be scrupulously fair in the execution of the appointment. Problems can arise when caught in the crossfire of the parties' different agendas and objectives. It is important to ensure, if possible, a common contract is agreed and signed jointly by both sides and that the SJE takes the initiative in regard to how your role develops. Do not be pressurised by one side only into a particular way forward. They should agree between them what opinions are required. If unagreed, remember that this is not the SJE's problem. They will need to apply to the court to have it resolved.

Fees

Difficulties can occur over fees. The SJE should ensure that these are agreed as part of the contract in the agreed terms of engagement from the outset binding on both parties in equal shares. In case of real doubts or uncertainty, ask for payment "upfront" by reference to staged amounts. I have managed to impose such a system successfully in at least two appointments. Do ensure that any variations or extensions to the advice/opinions required are agreed in writing with both sides. Any corrections to the draft report should also be agreed by both sides.

Any subsequent correspondence or questions of clarification within CPR Part 35.6 must be copied to the other side. Ensure that they have the opportunity to comment in advance of producing your responses to both parties. Any conferences or meetings must be attended by both parties. Do keep the court informed of any difficulties and the programme agreed for giving the evidence.

I have never given evidence in court as an SJE. However, it is believed that the SJE can expect to be asked questions by both sides. It is to be hoped that the judge will be particularly supportive in the handling of any undue exceptional hostility, but this cannot be guaranteed!

My advice is generally to be somewhat circumspect when asked to take on SJE appointments, particularly in respect of major actions or those with undue complexity. However, if an appointment is taken, bear in mind the following summarised points.

- **Instructions**

  At the outset, take the initiative to impose a common jointly signed appointment letter with conditions of engagement on both sides. The letter should contain precise instructions, including provisions for procedures, questions and a timetable for the various stages envisaged (this will satisfy the obligations in CPR Part 35.8 (2) to keep the other party informed).

- **Fees**

  Incorporate a specific fee agreement on a time basis within the instruction/appointment letter payable on a 50:50 stage basis if required (ie prior to reporting, dealing with questions, following the trial etc). If the SJE is uncertain as to risk, ask for fees "upfront", at least in respect of the pre-reporting stage. In case of payment difficulties, the court can give directions in regard to payment (CPR Part 35.8 (3)), but the court can also limit the amount of fees to be paid in advance of being instructed within CPR Part 35.8 (3) (a), but that amount may have to be paid into court.

- **Even-handed policy**

  The SJE needs to maintain a deliberate even-handed policy throughout. Never respond to any correspondence without consulting and copying the other side. Send a draft of responses to both sides seeking comments before signing off.

- **Meetings**

  Only attend meetings or pre-trial discussions when both sides are present.

- **Telephone calls**

  Do not answer any telephone calls with only one party. Specify in the instruction letter that such calls must be on a conference-basis only.

- **Communication with the court**

  Keep the court informed of progress at all stages, including CPR Part 35.6 questions.

- **Application to the court**
  In case of particular difficulties, inform the court and possibly consider making an application under CPR Part 35.14 (Part 35 rules are in appendix 1).

The Liability of an Expert Witness (Immunity)

The expert witness, when acting an as expert, usually enjoys "immunity from suit". This is a well-established principle of English law and, despite contemporary pressures, there have been (thus far) no successful attempts to upset the protection enjoyed as such.

The object of the immunity is to ensure that expert evidence can be given free reign of expression in the courts (or free, frank and open pre-trial discussions) without fear or threat of action being taken in any context for alleged mistakes or negligence. The primary duty is, in any event, to the court, not to the parties. The extent of the protection afforded was confirmed in the Court of Appeal judgment in *Stanton* v *Brian Callaghan & Associates* CA [2002] QB 75.

However, it must be stressed that the immunity does not extend to the expert when acting as adviser. Where, for example, advice is given as to the merits of a case, the normal potential liability for any breach of duty of care will prevail. Any failure to advise competently could precipitate a negligence claim for damages.

Also, it should be remembered that any untruths given under oath (which include any deliberate lies made in an expert's report) are subject to the usual risks of perjury being committed, which is a criminal offence.

The immunity extends to pre-hearing work (acting as an expert). This was confirmed in *Raiss* v *Paimano* [2001] Lloyds Rep 341, even when the expert was alleged to have acted dishonestly. The expert was not entirely honest about his qualifications and experience in his report and at the trial. At first instance, it was decided that the lies over qualifications did not come within immunity. However, subsequently the appeal judge disagreed.

It would appear that even where the expert is proved to have acted impartially (as in *Anglo Group plc* v *Winter Brown & Co Ltd*, referred to in chapter 3), no claim could be made. This situation has given rise to considerable debate. Where an expert misbehaves to the extent of impartiality, he may thereby have caused considerable financial loss to a party who may have relied on his evidence in the prosecution of his claim (or defence). However, it would seem that no compensation remedy will be forthcoming. The expert may (if he is professionally qualified) be reported to the disciplinary committee of his professional institution, but that is all.

The same immunity from suit applies in the case of the single joint expert (SJE), notwithstanding that the evidence of the SJE will be only available to the judge. If, therefore, it turns out to be wrong or unreliable, the judge will either have to determine the case without such evidence or, more likely, in the case of a highly technical case, adjourn while a replacement is found, thus adding to the costs.

Wasted costs

Solicitors and even barristers do not have total immunity before the court. They can also be the subject of applications for "wasted cost orders", for example where inexcusable delays have been caused. It had been widely believed that experts will not face such penalties, even where the court has been misled or time is wasted, for example because of inappropriate or impartial evidence.

However, in a recent case, *Phillips* v *Symes* [2004] EWHC 2330, this was held to be wrong on the facts of that case. A doctor's evidence as to the mental capacity of a claimant was strongly criticised by the court. It had been alleged that he had acted as an advocate giving biased evidence in breach of his duty to the court. The judge determined that, in exceptional cases, a cost order could be made against an expert witness who had been guilty of "gross dereliction of duty". He should not be able to rely on immunity in these circumstances. So beware. A further reason for compliance with the duty to act impartially.

Despite the luxury of apparent immunity, do not be lulled into a false sense of security. The law may change further and there is always a fine dividing line between evidence (from you as an expert) and advice (from you as an adviser).

Therefore, ensure that you have adequate professional indemnity insurance cover and be prepared to discuss with your instructing solicitor or client the extent of the cover that should be allowed. You may be asked to confirm this in your conditions of engagement.

Finally, prepare the independent and unbiased expert opinion with the same degree of gravity and care as if it were open to actions for negligence.

Privilege

15

In earlier chapters, I have referred to the protection afforded by privilege and which may be available as a safeguard against the disclosure of documents or other communications in court. Therefore, it is a matter of some importance as far as the expert is concerned, particularly when acting as an adviser. The expert will wish to have the comfort of knowing, for example, that documents and correspondence passing between him and his instructing solicitor fall within the protection. It is also vital to have this protection during negotiations or discussions with the other side under the safety net of "without prejudice".

Unfortunately, as a result of recent case-law, the seeming well-established doctrine in law of legal advice privilege may be undergoing some revision. There would appear to be a trend towards the greater restriction of the privilege concept, perhaps in reaction to the philosophy of greater disclosure and transparency in civil actions and other forms of litigation.

Privilege is considered under two distinct headings: the relatively well-understood protection of "without prejudice"; and the, perhaps, more esoteric principles of legal professional privilege covering both legal advice and litigation.

Without prejudice

The "without prejudice" principle will be familiar to many surveyors and valuers in the context of negotiations and discussions on value disputes, which could, if not settled, proceed to some form of litigation. The obvious aim is to promote the settlement of disputes by removing any fear that comments or offers made in correspondence could result in the party being compromised by its disclosure in subsequent legal proceedings. The operation of the rule aims to promote the optimum conditions for settling disputes and will apply where there are genuine settlement negotiations.

The "without prejudice" rule may be implied as a matter of law but, more commonly, arises out of the implied or express agreement of the parties. Thus, when parties agree to correspond in correspondence headed "without prejudice", they have confirmed that express agreement. However, it may be implied where there are settlement negotiations between parties. It is best to ensure that "without prejudice" cover is expressly applied when negotiating by correspondence, since this will ensure that the protection exists. In those circumstances, the court is unlikely to examine whether the negotiations were genuine (unless invited to do so). It is also always advisable at meetings to confirm verbally the assumption that all discussions are within the protection.

If one of the parties heads his initial letter "without prejudice", then the line of correspondence ensuing will likewise be protected unless one or both parties seeks to open by an express statement or agreement. On occasion, the protection will be partially opened as a result of what is known as a *"Calderbank* offer", where a party makes an offer to settle "without prejudice save as to costs". In these circumstances, the offer made can be produced to the court, tribunal or arbitrator but only in relation to the costs issues, but not before the previous judgment or arbitration award.

In *Reed Executive plc* v *Reed Business Information Ltd* [2004] EWCA Civ 159, it was held that the parties could not be compelled to disclose the details of "without prejudice" negotiations and this principle was unaffected by the defendant's refusal to take part in ADR and in relation to a costs issue (they had not Calderbanked).

Without prejudice protection cannot be subsequently waived by one party alone. However, the protection may only relate to the contents of the communication, not necessarily the fact that the offer was made. It has been determined, by the courts, that the "without prejudice" protection cannot be employed in the case of perjury, blackmail or what was referred to in *Unilever plc* v *Proctor & Gamble Company* [2000] 1 WLR 2436 as "unambiguous impropriety". This cumbersome phrase relates to an "unambiguous admission of facts" followed by an equally "unambiguous denial of those facts". In plain words, reneging on an agreement.

Legal professional privilege

Legal professional privilege can be subdivided into two distinct areas:

- Legal advice privilege.
- Litigation privilege.

Legal advice privilege

This privilege is in respect of confidential correspondence or other communications between a client and his lawyer, but only for the purposes of seeking or giving legal advice. Therefore, it would not protect any communication between the client or his lawyer and a third party, such as an expert witness. It could apply to an in-house lawyer as well as an external one so far as performing legal advice duties are concerned.

A major landmark case before the courts has recently dominated the subject of legal advice privilege. In *Three Rivers District Council v Bank of England* [2004] QB 916, the liquidators of the collapsed Bank of Commerce and Credit International SA (BCCI) were seeking disclosure of internal memoranda created within the Bank of England for the purposes of making submissions to The Bingham Inquiry set up by the Government to investigate the Bank of England's supervision of BCCI. The Bank had established a committee called the BIU to communicate with the Bank's lawyers.

The liquidators claimed that there had to be disclosure of internal memorandum being prepared by Bank staff for submission to the inquiry. The Bank claimed privilege for these documents. The liquidators had argued, successfully, that the BIU was "the client" and that only its communications with lawyers were privileged. As a result, documents prepared by Bank employees (not the BIU) were not privileged and could be produced. This was the judgment of the Court of Appeal upholding the decision of the High Court.

The judgment of the Court of Appeal reflected the ongoing inherent conflict and tension that exists in relation to legal advice privilege, the fundamental freedom guaranteed by the rule of law for a person or body to be advised of legal rights without disclosure to the world at large. However, this may undermine the public interest principles of maximising the open disclosure of facts in the courts.

It is, therefore, no surprise that, given the important constitutional issues at stake, permission was ultimately given (having been refused in an earlier parallel action) to appeal this judgment to the House of

Lords with the Law Society, the Bar Council and the Attorney-General intervening.

The House of Lords' decision goes almost all of the way to reaffirm the previously understood principles of legal advice privilege, thus overturning the decision of the Court of Appeal.

Lord Scott held that "presentation" advice or assistance given by lawyers to parties whose conduct may be the subject of criticisms by an inquiry is advice or assistance that would generally qualify for legal advice privilege. Legal advice was not solely in relation to advising a client on the relevant law; it may also include advice as to what should be done within the relevant circumstances. However, Lord Scott emphasised that it was necessary to identify the relevant legal context or, as another judge commented, "whether lawyers are being asked to put on their legal spectacles".

If, according to Lord Scott, the lawyer becomes in effect a general adviser on business to the client, the protection may not be available. It had been hoped that the Lords would clarify whether communication between an employer and his employees could be treated as falling within legal advice privilege but, unfortunately, they did not need to address this issue and therefore declined to do so.

Litigation privilege

It is emphasised that the House of Lords decision is only of direct application to legal advice privilege, not litigation privilege. However, Lord Scott indicated that while advice privilege may be safely assumed in the circumstances described by the judge, litigation privilege may no longer be. In comments, probably *obiter*, he inclined to the view that since litigation was no longer purely adversarial, a new look at litigation privilege was warranted but "that is for another day".

This comment has caused quite a stir in legal circles, given that most lawyers regard litigation as fundamentally "adversarial" although, as described earlier, veering somewhat towards the "inquisitional".

Privilege is in respect of documents that are brought into existence with the express purpose of being used in litigation and which are in existence or contemplation. They must be distinguished from documents or other evidence that are disclosed or found in the litigation process. Thus, an expert's report, correspondence and other advice to the client or his lawyers had been assumed to fall within the safety net. However, some documents that are obtained from third parties for the purposes of evidence in the legal proceedings may come within the privilege, but even this is doubtful.

Privilege may cover communications between a client or his lawyer and a third party, who are in aid of actual or contemplated litigation. No such protection ever existed in relation to correspondence or reports, which are passed, for example, between a surveyor and his client whether or not in contemplation of litigation. In this context, the often suggested manoeuvre of passing any such correspondence through the client's lawyer is unlikely to be sufficient, whatever the circumstances, to attract the protection, particularly if the lawyer had not been previously instructed to advise in contemplation of the litigation concerned. The intention is to get the benefit of legal professional privilege but that requires a need for legal advice.

Where does this leave the expert witness? Clearly, the protection historically assumed has been thrown into some doubt by the comments of Lord Scott in the *Three Rivers* case. As the law stands, litigation privilege exists. But the warning shots have been fired by the highest court in the land.

It would be as well for the expert to check with the legal adviser to whom a report is to be made if there is any doubt as to the likely extent of the privilege in any particularly case. Some lawyers will view with concern the suggestion that litigation under CPR is less adversarial than previously, although this certainly seems to be the case.

Experts will also wish to be advised as to whether a report made to their client that is only later disclosed in the litigation (as an expert's report for purposes of a trial) is likely to have been privileged.

In a recent case, *Jackson* v *Marley Davenport Ltd* [2004] 1 WLR 2926, this was the point at issue before the Court of Appeal. It will be recalled that CPR Part 35.13 provides that "a party who fails to disclose an expert's report may not use the report at the trial or call the expert to give evidence orally unless the court gives permission".

The court decided that the CPR were never intended to override the privilege enjoyed in relation to earlier drafts or reports from an expert who will, ultimately, give evidence if required at trial. CPR Part 35.13 merely confirms that a report cannot be relied on at trial if not first disclosed. That was the sole purpose.

However, in the later Court of Appeal decision in *Hajigeorgiou* v *Vasiliou* [2005] EWCA Civ 236, it was held that where the permission of the court was required to inspect an alternative second expert, the imposition of a condition requiring disclosure did not breach legal professional privilege and was therefore admissible.

This judgment will be of some comfort on this particular point. No mention was made of any loss, in any event, of rights of litigation privilege, but it was determined prior to the *Three Rivers* House of Lords' decision.

Arbitration and Independent Expert

Most commercial property valuers will be familiar with the almost universal adoption in commercial property lease rent review clauses of a reference to arbitration or to an independent expert in order to determine the rent on review if, in the event, the parties are unable to agree.

Background

The usual form of words in a rent review clause, coupled sometimes with other lease clauses, will be for the parties to agree the appointment of an arbitrator or expert. If they are unable to agree on the third-party identity, there is usually provision for the appointment to be made by application to the president of the RICS.

Thus, a substantial dispute-resolution industry within the profession has evolved to give effect to these third-party referrals. The dispute resolution service of the RICS advises the President on appointments and maintains a "panel" of suitably qualified and experienced arbitrators and independent experts covering the whole country and dealing with specific sectors of commercial property for this purpose. The panel of those who are commonly appointed is published on the RICS website, www.drs@rics.org.

Those who are appointed, in the case of both arbitrators and independent experts, have undergone the training courses administered by the RICS. In the case of arbitrators, this will include the production of a reasoned award. All those who are appointed are revetted every five years and have to satisfy the RICS as to continuing experience and CPD (continuing professional development) attendance at courses and seminars. Many senior arbitrators and independent experts have considerable knowledge of the relevant rent review law and are experienced as experts witnesses in their own right.

All arbitrators and independent experts appointed by the president are entitled to join Arbrix, which holds bi-annual conferences that provide legal practice updates.

While the majority of appointments are in respect of rent review disputes, I have, in the past, been appointed on other references, both privately and through the president, in relation to capital-value disputes where there is an arbitration or independent expert default provision. For example, where parties cannot agree the contractual sharing of the proceeds of a joint-venture property development.

It follows that commercial property experts will find themselves giving evidence to surveyor arbitrators or independent experts. Occasionally, they will also be required to give evidence before non-surveyor arbitrators, where alternative dispute resolution by this method is preferred. In the past, I have given evidence before judge arbitrators as to the value of property assets in a disputed company takeover and in disputed share allocations within a corporate property transaction.

A further area for possible arbitration references is the Property Arbitration on Court Terms (PACT) scheme devised jointly by the Law Society and the RICS, where the parties wish to proceed by a reference to arbitration in the case of lease renewal, rather than in the High Court or county court.

Arbitration

The practice of arbitration following an agreed reference by the parties as an alternative to an action in the courts has been long established. Generally, it has the merits of being private and confidential between the parties, quick, hopefully less expensive and, in the case of complex, technical disputes, decided by a professional person or persons having the relevant expertise and experience.

As a result of the development of arbitration practice in the UK, London became the preferred "seat" for many large, international trade disputes, such as in the shipping and commodity markets. Unfortunately, it became clear over time, that the previous Arbitration Acts 1950–79 (if adopted) were insufficiently drafted to give certainty and privity to arbitration proceedings. All too often the process was open to legal challenge and intervention, which threatened to derail the whole purpose and advantages of the arbitration process.

Arbitration Act 1996

A new arbitration bill was proposed and eventually this became the Arbitration Act 1996. It partially adopted the "model law" drafted by UNCITRAL (International Trade Committee of the United Nations) but retained much of English law where preferable, but in a much more serviceable mode.

As an arbitrator of some years standing, I do not know of any other Act of Parliament in recent years that has been so successful. This is particularly so in the rent review sector. It has given effect to a generally inexpensive, efficient and fair method of resolving such disputes. It enables the parties to determine (if they so wish and can agree) how the arbitration is to be conducted, but with certainty that the process will not be susceptible to unnecessary legal intervention. It is a good blend of the adversarial and inquisitional elements referred to earlier.

Except in the most major rent review arbitrations, where a hearing may be held with a solicitor or counsel representing the parties, the rent review surveyor acting as an expert providing evidence to the arbitrator will also be responsible for conducting the case. He will be involving himself in the conduct of the proceedings and may act additionally as an advocate. Therefore, it is essential to have a reasonable working knowledge of the Arbitration Act 1996, its powers, limitations, duties and obligations. Set out below are the principal provisions that will inform the expert on how to proceed to best advantage as far as his client is concerned.

Party autonomy — Section 34

The principles of Section 34 follow the model law already referred to but reverse those rights so that the arbitrator decides most procedures unless the parties otherwise agree. The agreement must be in writing in accordance with Section 5.

This is at the heart of the rationale. It is the ability of the parties by agreement to decide how they wish the arbitration reference to be conducted and affects most of the evidential and procedural provisions. The Act provides that "the tribunal" (the arbitrator) is to decide all procedural and evidential matters but subject to the right of the parties to agree any matter (in writing). The only exceptions are the mandatory provisions set out in Schedule 1, the most important of which in practice are listed below:

- Power of court to extend agreed time limits.
- Stay of legal proceedings.
- Power of court to remove an arbitrator.
- Liability of parties for fees and expenses of the arbitrator.
- Immunity of the arbitrator.
- Objections to substantive jurisdiction of tribunal.
- Determination of preliminary point of jurisdiction.
- General duty of the tribunal.
- General duty of the parties.
- Power to withhold award in case of non-payment.
- Enforcement of the award.
- Challenging the award.

As a rent review arbitrator, it has always surprised me that so little opportunities are taken by legally unrepresented parties to agree how they wish the dispute to be conducted. The most common exception is to exclude the Section 34 (g) right of the arbitrator to "take the initiative in ascertaining the facts and the law". In other words, the ability beyond the purely inquisitorial of making his own enquiries. Please note, however, that this does not prevent the arbitrator from asking questions of a party in order to clarify the evidence that has been submitted.

Jurisdiction — Sections 30, 31 and 32

The previous Court of Appeal decision in *Harbour Assurance Company Ltd* v *Kansa General International Insurance Company Ltd* [1993] QB 701 has been followed in the Act, enabling the arbitrator to rule on his own jurisdiction (unless the parties otherwise agree). This was a particularly welcome provision since, in my experience, much time and cost was expended previously in such disputes. Section 30 states that the arbitrator may rule on his own "substantive jurisdiction" as to:

- Whether there is a valid arbitration agreement.
- Whether the tribunal is properly constituted.

- What matters have been submitted to arbitration in accordance with the arbitration agreement.

Any ruling is open to challenge by application to the court, but only in restricted circumstances where the parties both consent or the arbitrator gives consent. The court must be satisfied that a substantial saving in costs will be made, the application was made without delay and there is good reason why the court should determine the matter (Sections 31 and 32).

Legal advice should be sought at the earliest opportunity. The options for challenge are not straightforward, particularly given the above preconditions. It should be noted that any objection on the grounds that the arbitrator lacks substantive jurisdiction must be raised by a party at the time when the first step in the proceedings is taken and must otherwise be raised as soon as possible after any jurisdiction point is raised.

A party that continues to take part in arbitration proceedings long after it could have raised any jurisdiction or fundamental procedural objections may lose the right. Therefore, for example, any objections raised late in the proceedings that the arbitrator was improperly appointed or gave improper directions or other such irregularities may be too late (Section 73).

(a) Duties of the arbitrator and the parties — Sections 33 and 40

Sections 33 (1) and (2) set out the arbitrator's obligations in accordance with the rules of natural justice. The intention was to move away from any preconceived court-like procedures, and merely to lay down general rules. The obligations are important since they may determine whether the arbitrator may have been guilty of "serious irregularity", which is one of the two grounds for challenging an award. The section states the tribunal shall:

- Act fairly and impartially as between the parties, giving each party a reasonable opportunity of putting his case and dealing with that of his opponent.
- Adopt procedures suitable to the circumstances of the particular case, avoiding unnecessary delay or expense so as to provide a fair means for the resolution of the matters falling to be determined.

- Comply with that general duty in conducting the arbitral proceedings in its decision on matters of procedure and evidence and in the exercise of all other powers conferred on it.

The general duty of the parties is set out in Section 40 as follows:

- The parties should do all things necessary for the proper and expeditious conduct of the arbitral proceedings, which include:
  - complying without delay with any determination of the tribunal as to procedural or evidential matters, or with any order or directions of the tribunal
  - where appropriate, taking, without delay, any necessary steps to obtain a decision of the court on a preliminary question of jurisdiction or law.

Default — Sections 41 and 42

The Section 40 provisions widen the responsibility originally set out in the Arbitration Act 1950. It is important in the case of a defaulting party, since the arbitrator will have usually provided directions and non-compliance would be a breach of Section 40. If the breach continues, the arbitrator may need to make a "pre-emptory" order, otherwise known as an "unless order". In the case of a continuing default following an "unless order", the arbitrator may then have no option but to proceed *ex-parte* (ie without the involvement of the party in default). For example, where a defaulting party does not participate in providing evidence as directed.

The specific powers of the arbitrator in the event of continuing default are in Section 41 and include a preamble of remedies in the case of "inordinate" delay by the claimant or the failure of either party to comply with the procedural directions. The remedies include dismissing the claim (in the case of the claimant) or any number of actions following the failure to comply with the pre-emptory ("unless") order, such as proceeding *ex-parte* or making an order as to costs. Section 42 enables a court order to be obtained requiring a party to comply with a peremptory order by the arbitrator.

(a) The fees and costs of the arbitrator — Section 28

Under Section 28 (1), the parties are jointly and severally liable to pay for the arbitrator's reasonable fees and expenses (if any) as are appropriate in the circumstances. It will be usual for the arbitrator to set out the basis of his fees at the outset and, whether agreed or not, it will usually be included in directions.

I find that surveyors representing clients in rent review disputes do not often fully understand the full impact of these provisions. If there is a dispute as to the amount of the fees, either party may apply to the court (for a settlement). However, where agreed or determined, in the absence of any arrangement between the parties as to their respective liabilities, they are jointly liable.

To take an example, where a landlord applies for the appointment of an arbitrator and, ultimately, the claim fails without an award being made (ie the landlord withdraws his demand for a review increase), it is not open to the lessee to reject any invoice for interim costs from the arbitrator. In the absence of any agreement between the parties, the parties are both liable for 50% of the arbitrator's costs. Therefore, it is vitally important to ensure that any settlement prior to an award covers the question of liability for the arbitrator's costs and fees. In the event of any dispute, it is still open for either or both parties to apply for an award of costs (Section 51).

As is probably well known, the arbitrator will not release his award until his fees have been paid, usually, at that stage on a 50:50 basis. The right to refuse delivery in the absence of payment is specifically given in Section 56 (1).

(b) Costs — Sections 61, 63 and 64

The Act defines what is meant by the "costs of the arbitration". It may include the arbitrator's fees and expenses, fees and expenses of the institution appointed (ie in rent review cases the RICS) and the legal or other costs of the parties. All may be included in any award, being the "recoverable costs" of the arbitration.

Some older leases provide that the costs of a rent review dispute at arbitration shall be shared in a particular fashion, whatever the outcome. Section 60 specifically provides that any such agreement will only be valid if made after the dispute has arisen. Thus, do not rely on any such provision when advising a client on an arbitration reference.

Assume that the usual rights to award of costs will apply, unless the parties have otherwise agreed after the dispute has arisen.

However, within Section 63 the parties can agree as to what costs in the arbitration specifically are recoverable.

In the absence of any specific agreement, the arbitrator may determine by award the recoverable costs of the arbitration "on such basis as it thinks fit". Therefore, he has complete discretion but should set out in his award the basis on which it has been determined, giving the items of recoverable costs involved.

While having a large measure of discretion, the arbitrator is directed in Section 61 (1) to award costs on the well-established principle that costs should "follow the event" except where this is "not appropriate".

Giving effect to this principle is the subject of much debate in arbitration circles. There are many possible permutations of treatment.

By its nature, a rent review award is a valuation determination somewhere between the respective positions taken in evidence by the parties. Thus, very often neither party will be wholly "successful". This can be distinguished from a non-rent review arbitration, for example, on a claim for damages in an alleged breach of contract. Where such a claim succeeds (even only to a small amount), it will be seen as justifying the action in the first place. In such a situation, costs will often be wholly awarded against a respondent. However, in a rent dispute, there may be elements of justification on both sides. Thus, the practice of "fractional" costs awards has developed.

Where a *Calderbank* offer (an offer made to settle the review "without prejudice save as to costs") has been made, the position will usually be different if the party making the offer was wholly successful. For example, if the tenant made a *Calderbank* offer to settle the sum higher than the arbitrator's award, it will be usual for it to be awarded all of his costs against the landlord. The reverse will also be true.

Arbitrators have to make potentially difficult decisions in the case of a "near miss". By this, what is meant is a situation where the landlord or tenant's offer was marginally higher or lower than the award. It would seem in accordance with natural justice that being adrift by only a nominal amount should be viewed as if a wholly successful *Calderbank* offer had been made.

It will be appreciated that in a costs award (unless otherwise agreed by the parties) the arbitrator will have the ability to award on all costs, including those of the costs award itself.

If a *Calderbank* is successful, generally the costs discretion in that respect will only date from the last day upon which the *Calderbank* could have been accepted by the other party. The arbitrator is under a duty to spell out in detail how the costs are allocated as between the parties in his award. In any submissions on costs, it is important that the full details of what is claimed to be recoverable from the other side are shown under appropriate headings.

Within Section 65, the arbitrator has the power to limit the recoverable costs and thus may limit the amount awarded to any party. Thus, the arbitrator can make an order to this effect at any stage, usually, in practice, as a result of an application by the other side. This may be appropriate if, for example, one of the parties employs expensive legal representation, has expensive witnesses or employs tactics during the arbitration proceedings that incur excessive, unnecessary or unreasonable costs.

This is in tune with the doctrine of "proportionality" — one of the cornerstones of CPR. In practice, it is usually only sufficient to warn an offending party of the arbitrator's ability to make such an order where excessive costs are in prospect.

(c) Challenging the award — Sections 68 and 69

Generally, the courts are reluctant to intervene in arbitration awards and proceedings, particularly since the advent of the Arbitration Act 1996. The only grounds for challenging the award are to be found in Sections 68 and 69.

Section 68 concerns the question of "serious irregularity" and is a welcome change from the previous unfortunate description of "misconduct". The section specifies nine categories of possible failure by the arbitrator that could be deemed to be serious irregularities "which has caused or will cause substantial injustice to the applicant". They are:

- Failure to comply with the general duty (Section 33).
- Exceeding the powers (ie jurisdiction).
- Failure to conduct proceedings as agreed (eg in directions).
- Failure to deal with all issues.
- Exceeding the powers.
- Uncertainty or ambiguity as to effect of award.
- Award obtained by fraud or in breach of "public policy".
- Failure to comply with requirements (as to form of an award).
- Any admitted irregularity in proceedings or award.

In such circumstances, the court may remit the award back to the arbitrator for reconsideration, set the award aside or declare it to be of no effect, but only where "injustice" has been or will be caused. In a recent case in *Margulead Ltd* v *Exide Technology* [2004] 2 All ER 727, the issue was whether an award should be set aside because of a "serious irregularity". The arbitrator was said to have failed to allow a right of reply orally to closing submissions by Exide's counsel and a failure to refer to an agreement put by Margulead. On the particular facts, both grounds were dismissed. It appears that Margulead failed to raise these issues at the time of the hearing. This judgment emphasised the restricted approach of the courts to challenging an award and the need to raise such an issue at an early stage (preferably at the hearing).

However, in another recent case, *Vee Networks Ltd* v *Econet Wireless International Ltd* [2005] 1 All ER 303, the High Court confirmed that:

- A party wishing to challenge substantive jurisdiction had to do so at the start of the proceedings (see *supra*). The Section 67 claim failed.
- The arbitrator's clarification into statutory construction without giving the parties opportunity to comment was a Section 68 irregularity. The court did not need to determine whether substantial injustice had been caused only that there had been an irregularity of procedure, which caused an unfavourable conclusion to the application.

The award was remitted for reconsideration.

The alternative basis for a possible challenge is under Section 69 on a question of law arising out of an award. Leave to appeal will only be given if the court is satisfied that:

- The determination of the question will substantially affect the rights of one or more of the parties.
- The question is one that the tribunal was asked to determine.
- The decision of the tribunal on the question is obviously wrong.
- The question is one of general public importance and the decision of the tribunal is at least open to serious doubt.
- Despite the agreement to resolve by arbitration, it is just and proper in all the circumstances for the court to determine the question.

Decided cases illustrate a general reluctance by the courts to intervene. Where successful, it has usually been the case that the award has been remitted to the arbitrator for reconsideration with directions to correct any irregularity or to award on a different interpretation of the relevant law. Generally, it has been determined that, in considering arbitrators' awards, there is a presumption of finality, which may nevertheless be rebutted on the facts in the light of the above tests.

In a recent case before the High Court, *St George Investment Co* v *Gemini Consulting Ltd* [2005] 01 EG 96, the judge remitted for reconsideration to the arbitrator an award in an office rent review because of serious irregularity. The arbitrator had introduced a specific rental discount for certain features of the lease that had not been the subject of the evidence put to him by the parties' expert valuers. He had departed from the agreed basis upon which the case had been submitted.

A former colleague acted as an arbitrator in a rent review dispute. Although his award was comprehensive, it omitted to address one of the comparables put by one of the parties. This was deemed to be technically "a serious irregularity" and the award was remitted to him for reconsideration. He revised his award incorporating in his "findings and reasons" comments on the missed comparable, which had no impact whatsoever on his award. A pyrrhic victory for the appellant landlord!

(d) Reasoned award — Section 52

The award will be reasoned unless an agreed award or the parties have agreed to dispense. The award must, for legal reasons, state the "seat", which in most cases will be England and Wales.

There are no directions otherwise in the Act as to precisely how an award is drafted. However, arbitrators generally follow a predetermined structure, some being rather more comprehensive in their findings and reasoning than others. There is a public duty to give sufficiently comprehensive findings and reasons so that the parties can establish, with reasonable accuracy, how much weight has been attached to the evidence. Thereby, the parties will have an understanding as to how the various issues have been determined.

The question of when irregularity may have occurred is, of course, highly subjective. However, it is suggested that where a reasoned award does not address any or all comparables (either sufficiently or at all) selected by each party or other major issues and/or legal or other substantive points raised, there may be grounds for an appeal. Serious irregularity may also have occurred, as in the *St George*

Investment Co v *Gemini Consulting* [2005] 01 EG 96 case, where the arbitrator has used his own valuation skills outside the parameters of the evidence without seeking the parties' views. However, each reference will vary enormously according to the circumstances. Therefore, no hard or fast rules can be laid down.

Legal advice should be obtained if it is believed there to have been serious irregularities or a substantive legal point has been misconstrued. One of the most common errors, in my experience, is where a rent review arbitrator takes advantage of his inquisitional powers under Section 34 (9) "to ascertain the facts and the law" himself but fails to put his findings to the parties for their comments, thus breaching the duty to act "fairly" (Section 33 (1) (a)).

(e) Arbitrators' immunity — Section 29

Section 29 is short and to the point. An arbitrator is not liable for anything done or omitted in the discharge or purported discharge of his functions as arbitrator unless the act or omission is shown to have been in bad faith.

An arbitrator who fails to act at all is probably in breach of contract and may be open to an action on those grounds. Section 29 is only concerned with failure to act properly in the conduct of the arbitration. The Section 29 immunity does not affect any liability incurred by an arbitrator by reason of having to resign.

Clearly, questions of bad faith go beyond matters of serious irregularity where the arbitrator may, nevertheless, have believed that he had acted properly. It is suggested that bad faith may include such matters as fraud, dishonest or demonstrable bias against one party, for whatever reason.

(f) Removal — Sections 23 and 24

Only the parties acting jointly or the arbitral institution appointing (in the case of rent reviews, the president of the RICS) can revoke the authority of an arbitrator (unless the parties have otherwise agreed).

Within the powers given in Section 24, either party may apply to the court for the removal of an arbitrator on any of the following grounds:

- Circumstances exist that give rise to justifiable doubt as to his impartiality.

- He does not possess the qualifications required by the arbitration agreement.
- He is physically or mentally incapable of conducting the proceedings or there are justifiable doubts as to his capacity to do so.
- He has refused or failed to conduct the proceedings properly or to use all reasonable despatch in conducting the proceedings or making an award and that substantial injustice has been, or will be, caused to the applicant.

Unless mutually agreed by all participants, the grounds upon which an arbitrator can be removed are therefore restricted. An application to the court is necessary. The most common grounds are likely to be under the first or last point above. In my experience, the former is the most common.

Occasionally in the course of the arbitration, it becomes apparent that the arbitrator has a potential conflict of interest, which had not been previously disclosed either before or after his appointment. Note that only "justifiable doubts" are needed in this context, but it is suggested that the courts will, again, be unwilling to intervene unless the risk of obtaining natural justice is a valid concern.

However, in considering any possible grounds as to "justifiable doubts", it is worth having regard to the judgment in *Laker Airways Inc v FLS Aerospace Ltd* [2000] 1 WLR 113. The case concerned Mr Burton, a QC appointed as arbitrator, whose colleague in the same chambers was acting for FLS. Laker applied to the court for the removal of Mr Burton under Section 24 (1) (a). Mr Justice Rix ruled that the circumstances did not give rise to doubts as to impartiality, but in the course of so finding gave some useful guidance on the subject. The test was an objective one and would rely on any one of the following:

- Actual bias (ie on the facts).
- Real danger of bias (note the word danger rather than likelihood).

It is of note that the last two points are wide in their potential catchment, particularly since, objectively, they must be viewed from the standpoint of third-party perception not the arbitrator's own belief and conscience.

(g) The preliminary meeting

Except in the simplest or more straightforward cases, the arbitrator will usually request or order that there is a preliminary meeting. In my experience, it is advisable to have one if possible. It will enable the parties to have their say on precisely how the arbitration is to be conducted. In practice, much time will be saved by a full review of the overall position of the reference. It is suggested that the valuer attends with a clear agenda from the client of what matters are to be impressed upon the arbitrator (other than the evidence itself, which will not be considered since the meeting is purely for procedural purposes). The arbitrator will not be impressed if there is an attempt to influence the possible outcome at this juncture, but there is the opportunity to apply in respect of how the evidence is to be adduced and on what basis.

The arbitrator will wish to know the nature of the dispute. In rent review cases, this takes on less significance since the arbitrator will have seen the lease and it will probably be obvious that the determination of the "open market rent", or similarly worded amount, is at issue. The award will be based primarily on expert valuation opinion relying on comparable evidence.

Unless agreed, the arbitrator will determine whether the evidence will be in written submissions or given at a hearing. In my experience, the majority of rent review arbitrations are conducted on a "documents-only" basis. Usually, the arbitrator will reserve the right to call for a hearing in the case of any evidential difficulties arising (unless the parties otherwise agree).

A preliminary meeting will take on particular significance for a party, where the other side prove to be dilatory in agreeing facts and otherwise responding to the dispute. Certainly any questions of default could be the subject of an application at this stage. In the event that one of the parties fails to attend, the arbitrator will normally adjourn and arrange a further meeting. If again there is no attendance, he may be minded to proceed *ex-parte* following the requisite default procedure, which may include an "unless" order.

This is also the time for the valuer to raise any questions in regard to the arbitrator's jurisdiction, fees and whether there are any substantive points of law involved in the dispute. In the latter event, there are alternative ways in which substantive legal points can be resolved. In the absence of a written agreement between the parties, it will be for the arbitrator ultimately to decide, but he will listen to the parties' views. The arbitrator can determine legal points as part of his award, decide the points as a preliminary issue or allow the parties to apply to the court.

The arbitrator will deal with the requisite rules of evidence and usually determine (unless the parties otherwise agree) how comparables or other evidence are to be adduced. Usually, this will involve the completion of a RICS-style rent review evidence pro-formas signed by a party, his agent or surveyor involved in the transaction, unless the parties agree the comparables in the statement of agreed facts or one of the surveyor experts was involved in the transaction.

Section 34 covers procedures and the questions of how evidence is to be delivered. Again, this is for the arbitrator to decide (unless the parties otherwise agree). The matters covered include:

- When and where any part of the proceedings is to be held (ie if a hearing).
- The language to be used.
- The form of statements of claim and response.
- The classes of documents to be produced.
- Questions to be put (when and in what form).
- Whether strict rules of evidence apply.
- Whether the arbitrator can take the initiative in ascertaining the facts and the law.
- Whether oral or written.

Normally, the arbitrator will require parties to agree in any event a statement of agreed facts and a statement of agreed comparables, which may incorporate rental evidence pro-formas.

In the case of a documents-only procedure, the arbitrator will discuss a timetable to be followed and there will be the opportunity to agree the most convenient dates. The arbitrator is under a duty to proceed with reasonable diligence (and so are the parties — Section 40) and, therefore, any lengthy periods proposed by one or both parties may be discouraged.

In regard to the question of fees, it is my experience that there is rarely any serious disagreement on this issue or in regard to the time basis that is usually agreed. In the event of a serious disagreement, the arbitrator may have no option to proceed in any event, but will then be subject to an application to the court.

In the unlikely event that the arbitrator proposes a fee basis that is related to the amount of the award, objections should be made on the grounds that this could be against natural justice and may indicate a bias in favour of the landlord. The valuer should not expect that the arbitrator will be prepared to "cap" his fee and it would be unwise to request this. However, he may be persuaded to provide a non-binding estimate or fees "budget" for the parties' guidance.

In the event that it is decided in the absence of agreement or otherwise that a hearing is held, different procedures need to be considered. It is suggested, to the uninitiated, that legal advice and probably (unless the sums involved do not warrant the additional cost) advocacy representation should be sought before embarking on the procedures involved. While the statements of agreed facts and comparables will sometimes be the same, usually parties will be asked to provide a statement of claim and response with the potential for sequential pleadings thereafter. Separately, it will be necessary to agree the dates when the experts' reports are to be exchanged. The arbitrator will also confirm the arrangements for the hearing and whether evidence is to be taken on oath. Parties will be asked to estimate the likely extent of the evidence and the number of days' hearing that may be required.

It is also at the preliminary meeting that any difficulties over witness disclosure of documents should be raised. The arbitrator has to give his permission for an application to the court (Section 43). In a rent review case, arrangements for disclosure between the parties are not normally significant but other issues may be involved. The arbitrator may wish to consider making orders in accordance with his powers granted under Section 34 as to the documents or class of documents that are to be disclosed between the parties and at what stage. (Unless the parties otherwise agree as to how they wish this to be treated.)

Third-party disclosure applications, needing approval by the arbitrator, are sometimes made in the case of rental evidence that is relevant to the case but cannot be released by the parties (or more often their agents) possibly because of a confidentiality agreement. An application (to the court) is invariably successful in unlocking such evidence by the issue of a witness summons under Section 43 (2). However, in a recent case, *South Tyneside Borough Council* v *Wickes Building Supplies Ltd* [2004] NPC 164, the court made an exception where the evidence was held by a competitor of Wickes (in this case B&Q) and was deemed to be commercially sensitive because both companies were bidding for the same site.

To summarise, the issues that will usually be on the agenda at a preliminary meeting in the case of a rent review reference are:

- The appointment and jurisdiction.
- The parties and other advisers/expert witnesses.
- Confirm nature of dispute/issues.
- The "seat" of the arbitration.
- Documents — the lease or leases — rent review clause.
- Procedures — hearing or written submissions/counter-submissions or written agreements?
- Section 34 written agreements.
- Evidence — rules, hearsay, admissibility, Civil Evidence Act 1995.
- Statement of agreed facts.
- Statement of agreed comparables.
- Disclosure, attendance of witnesses.
- Expert witnesses — *RICS Practice Statement*.
- Points of law (ie whether any likely).
- Privilege — "without prejudice" protection.
- Confidentiality.
- Timetable for agreement of facts, written submissions or statements of claim and response, exchange of experts' reports, counter-submissions (or statement of claim and response, if a hearing).
- Inspection arrangements.
- Fees and expenses (arbitrator) and when payable.
- Appointment of legal assessor.
- Costs — recoverable, limitations.
- Rent review award — reasons.
- Costs award.
- Applications.

In most rent review disputes, the preliminary meeting is the only occasion when the party's surveyor will have the opportunity to meet the arbitrator and discuss the way forward. The valuer should be as helpful as possible and assist the arbitrator in his task of setting down the procedures as far as is possible without departing from clients' instructions. If he wishes to make applications on any particular point, this should be carefully planned, giving substantive reasons and justifications, and always appearing to be reasonable in accordance with good advocacy.

If possible, it would be as well to iron out any potential difficulties with the other party's valuer prior to the preliminary meeting, including any written agreements on procedure or methods of adducing evidence.

Following the preliminary meeting, the arbitrator will issue his written directions. If it is intended to make an application on any issue, do this immediately rather than later in the proceedings when it may be difficult, or impossible, for the arbitrator to make any order that will favour the application.

(h) *Expert evidence advocacy and "spin"*

I have already referred, at some length, to this issue earlier related mainly to civil proceedings. No apology is made for reintroducing the debate in the context of arbitrations. It is emphasised, in my experience and that of most fellow arbitrators, that there are serious points of concern with regard to current rent review arbitration practice.

Unfortunately, I have to relate that of all the arbitrations I have handled over the years, very few have complied fully with the *RICS Practice Statement*. Some mix expert opinion with advocacy and, in addition, often exhibit elements of exaggeration, selectivity of evidence and bias. In a few cases, they were downright dishonest!

This matter has become of such great concern to the RICS Dispute Resolution Service and the majority of arbitrators that the RICS is contemplating revised rules, guidance and enforcement, which will post-date the publication of this book.

Conflicting opinions have been voiced as to how this might be rectified. Suggestions range from better education and training, to greater use of disciplinary powers through to only relying on submissions as advocacy. It should be emphasised that even if only viewed as advocacy elements of "spin", selectivity or exaggeration should not find a place. An advocate must never distort the facts.

After careful consideration of all factors, the majority of arbitrators have voiced their support in continuance of expert evidence to be provided in rent review arbitrations by the parties' surveyors. However, the surveyor representatives may additionally wish to apply their advocacy skills in persuading the arbitrator of their party's case, but this must be made separately from the expert opinion as such (in accordance with the *RICS Practice Statement and Guidance Note* — see chapter 4).

Unfortunately, I have encountered cases where the evidence presented by both parties was so unreliable that the arbitrator was placed in real difficulties and it was necessary to "step down into the arena" in terms of ascertaining the facts coupled with detailed questioning of the expert witness valuers in order to identify their "real" opinions. This should not be necessary. The rent review valuer should remember that overuse of inelegant advocacy, possibly coupled with highly exaggerated or selective evidence, is unlikely to assist the client's cause. Very little weight will be attached to evidence that is so coloured and may be revealed as unreliable under cross-examination or in counter-representations.

Whether counter-representations are, in any event, pure advocacy is debatable, but even here it is better merely to challenge the facts produced by the other side and give the reasons why the valuer's evidence is to be preferred. If more aggressive examination and rebuttal are required, then it should be made clear when this is being applied to at least part of the counter-representations.

The client should be made aware of the change of role and the reasons why (as expert witness) it is important to take the impartial objective route and why the fee arrangements (if contingency-based) may need adjustment (see *RICS Practice Statement and Guidance Note*).

Finally, comment is made upon two issues, which commonly arise when expert evidence is put to me as an arbitrator. Reference has been made earlier to the need for a "valuation trail" in preparing evidence before the court. The same principles are of equal importance in evidence to arbitrators.

It is often found, in practice, that the expert witness, having selected the most relevant comparable evidence, then fails to link the devalued and possibly adjusted rents or capital values to the method and basis upon which valuation opinion is produced.

It is not merely sufficient to refer to comparables X, Y and Z as being preferred, then to arrive at a valuation opinion made in the light of these. The thought processes and mathematical adjustments separating one from the other must be explained in detail. Do include the range of adjustments made to each comparable relating to such matters (in the case of office properties) as specification, location, age, service charge provision, conversion of headline rent to net effective and so on. The valuer cannot merely rely upon a bland expression of experience to justify his opinions.

The second important point concerns brevity and comparables. Lawyers are generally very good at narrowing a case down to the essential issues; surveyors less so. There is a tendency to include reference to all the comparable evidence or most of it. But, in my experience, quality comes before quantity. The valuer should select only the best comparables for inclusion in the valuation "trail". He may refer generally to others that are supportive but limit the sequential analysis to the few which opinion dictates should attract the greatest "weight".

Independent expert

Although not strictly a quasi-judicial process, the submission of evidence to an independent expert rather than an arbitrator falls within the *RICS Practice Statement*. It must give rise to the same considerations since expectations as to reliability will be the same.

On the subject of advocacy or expert opinion, the issue is believed to matter less in respect of submissions to an expert. It would be acceptable, to me as an expert, if only advocacy was employed, as long as it is honest and free from gross exaggeration but necessarily presenting the available evidence in the best light. However, the expert should be left in no doubt as to how the presentation of the case is being made.

While there are similarities in practice to the arbitration process, a reference to an independent expert is fundamentally different.

The independent expert is undertaking a valuation determination in accordance with the rent review clause and the directions given in the lease as opposed to an arbitrator, who is charged with the duty of arising at an award based primarily on the evidence put to him. The essential differences are sometimes misunderstood and it is important to make the distinctions.

Arbitration is a quasi-judicial process while the independent expert or surveyor appointed by both parties is undertaking a valuation role. It is certainly the case that the two roles appear to be drawing closer together in recent years partly by reason of the arbitrator's ability, under the Arbitration Act 1996, "to ascertain the fact and the law", which may involve valuation investigations (unless the parties otherwise agree). The independent expert will usually be receiving submissions of opinion from both sides. It would be a foolhardy expert who does not invite written opinions of value from both parties given his legal liability for the valuation opinion which he may produce, whether or not this is a requirement of the lease.

Despite the essential differences, rent review surveyors often stumble over the different procedures and sometimes expect me, as independent expert, to produce "directions" and anticipate that rules of evidence will be enforced. The independent expert has no such powers. He will usually set down a suggested procedure for agreement between the parties for the sake of good order and natural justice. However, it should be recorded that in the very recent Court of Appeal judgment in *Amex Civil Engineering Ltd* v *Secretary of State for Transport* [2005] EWCA Civ 291 it was held that an expert "decision maker" does not have to comply with the rules of natural justice. He has no quasi-judicial means of enforcing any agreed procedures, although it is believed that he could seek to enforce the "contract" with the parties in the event of serious default if both parties have signed up to the agreed procedures.

The independent expert will usually wish to maintain an objective stance throughout his appointment and will encourage the parties to adopt procedures on submission of evidence that do not prejudice his independence.

Since the independent expert is primarily using his own experience and expertise as a valuer in respect of the class of property in the relevant location, the suitability of his appointment essentially as a valuer may have to be judged by different criteria. Many leases prescribe special requirements for an appointment by the president of the RICS. Typically, words such as "the independent expert shall be a director or partner of a major local commercial practice who is experienced in the valuation and letting of similar type office premises".

Unfortunately, most rent review surveyors and valuers, at least in the major commercial surveying practices, are rarely involved on a day-to-day basis with the letting of commercial property. The significance of this has not yet made its impact upon some solicitors who draft commercial leases. The result is that the RICS is experiencing increasing difficulties in securing appointments, except where the parties are willing to accept a more liberal interpretation of the words "letting experience".

This issue was recently reviewed by the Court of Appeal in a case concerning a retail store in Jersey where an appointment was made by the president of the RICS. The valuer was to be "experienced in the valuation and letting of premises so far as practicable of similar character or comparable to the demised premises within the island of Jersey". The tenant failed in his application for wrongful appointment for reasons connected with the type of retail store, but the court did determine that the agreement between the parties and the lease was fundamental. The president should not appoint someone who does not comply with the terms so that the complainant party has to challenge the appointment as being invalid.

The judgment may lead to further difficulties in obtaining appointments in some cases and possibly uncertainties in regard to the challenge of experts' determinations where suitability is an issue, and, therefore, a challenge to the appointment was possible.

When an independent expert is appointed, it is important to check that the determination was made fully in accordance with the terms of the lease, which will be usually prescriptive. Normally, this will be a determination (by valuation) of the rent on review (often the "open market rent") within the terms of the rent review clause.

The Court of Appeal in *National Grid Company plc* v *M25 Group Ltd* (No 1) [1999] 1 EGLR 65 had before them the issue of a lease where the independent surveyor had to determine "the question of whether any or if so what increase ought to be made in the rent payable". The court determined that "the valuer must ascertain the rent in accordance with the contractual criteria, but he does not have a completely free hand in deciding the question of what increase ought to be made in the rent payable". The independent expert cannot expressly, or by implication, construe the terms of the lease (as maybe an arbitrator can).

Quite often in arbitration and independent expert appointments it is found that one or both parties ask me to determine the increase in rent payable, ignoring the substantive wording of the rent review clause, which asks me to award or determine "the open market rent" or similar. Of course, the latter may be above or below the previous "passing" rent but, in the case of arbitration, would normally be within the range of the expert valuations submitted. The expert should be certain to recite the exact wording of the clause in evidence and relate to his valuation.

The Lands Tribunal

17

The Lands Tribunal was established by the Lands Tribunal Act 1949 and started life on 1 January 1950. Its jurisdiction is in respect of all parts of the UK outside Scotland. It is an independent judicial body within the overall court service.

The then Attorney-General, Sir Hartley Shawcross, said that the object of the bill was to:

> Strengthen and codify the statutory arrangements for settling disputes in connection with the valuation of land and to ensure that, in all cases, where some form of valuation under statute is required there should be a single and consistent jurisdiction, combining legal and technical experts experience with an appeal on matters of legal importance.

The original intention was that the tribunal would replace the official arbitrators appointed under the Acquisition of Land (Assessment of Compensation) Act 1919 and under the Finance Act 1910 but, during the committee stage of the Lands Tribunal Bill, the tribunal was given the jurisdiction of hearing rating appeals from local valuation courts. This type of appeal has since represented a large proportion of the tribunal's work.

The Lands Tribunal may also act as an arbitrator by consent of the parties in the case of land valuation disputes, such as rent reviews and compensation claims. The work of the tribunal can now be summarised in the following areas:

- **Rating appeals**
  Rateable value disputes on appeal from what are now called the local Valuation Tribunals.

- **Leasehold enfranchisement**
  Appeals from the decisions of leasehold valuation tribunals (not covered in this book).

- **Restrictive covenants**
  Applications for the discharge or modification of restrictive covenants under Section 84 of the Law of Property Act 1925.

- **Compensation**
  Disputed compensation claims arising out of the compulsory acquisition of land (meaning land and buildings).

- **Tax valuations**
  Appeals against assessments made by the Inland Revenue under the Taxes Acts.

- **Rights of light and blight**
  Valuation disputes in respect of rights of light and blight notices under the Planning Acts.

The tribunal normally consists of a president, a former judge or barrister and other members consisting of both lawyers and surveyors experienced in the valuation of land. At the time of writing, there were three full-time chartered surveyor members. The majority of cases are heard by a single member. Major actions that involve substantial claims financially, or where there are significant issues of law, may be heard by a two-member tribunal consisting of a lawyer and surveyor.

Procedures

Since, in some cases, the expert witness surveyor or valuer will be involved in the administration of the case to the Lands Tribunal, brief reference is made to the procedures involved. However, except in the very simplest cases, it is recommended that a client's solicitor should deal with the necessary applications and compliance with the Lands Tribunal rules and procedures.

A written notice must be sent to the Registrar of the Lands Tribunal with all the necessary information and accompanied by a fee as prescribed in the Lands Tribunal explanatory documents. Once registered, a new case is allotted to the "standard" or "special procedure".

"Special procedure" is for cases of high value or complexity but, for most, the standard procedure is appropriate. In very small or simple cases, there is a "simplified procedure" with no legal representation required and the surveyor is allowed to act both as advocate and expert witness and the hearing is informal. There will be no award of cost in such cases.

There is, in addition, a "written representations procedure", where the tribunal will reach a decision based on documents only. It is for the tribunal to determine whether the case is suitable for this reduced procedure.

As in the courts, witnesses of fact or expert opinion will give evidence on oath. Chartered surveyor valuers making an appearance as an expert witness will need to comply with *RICS Practice Statement* as described in earlier chapters. Expert witness reports must include a statement of truth in accordance with the prescribed wording set out in Part 35 of the Civil Procedure Rules 1998 as in the case of the High Court. Under CPR, expert witnesses have the same duty to the tribunal as they do to the courts.

Rating appeals

An appeal from a local Valuation Tribunal must be lodged with the registrar of the Lands Tribunal within 28 days of the Valuation Tribunal decision. At present, the fee payable is 1% of the rateable value (minimum £50 and maximum £5000). The other party to the appeal must indicate within 28 days if it wishes to be the respondent. In a further 28 days, the appellant submits a statement of case and the respondent makes his reply within a further 28 days. These time limits can be extended by application and where expert evidence is to be submitted (as it will be in the majority of cases), these will need to be lodged together with supporting documents.

The parties are expected to bring to the hearing documents, plans, valuation reports, legal precedents and such like that are to be relied upon in the hearing.

Any award of costs will as in arbitration "follow the event". Exceptions will be made where the appellant achieves only a minimal variation in the determination of the tribunal, or where the conduct of the successful party has unnecessarily increased the costs of the unsuccessful party.

Compensation references

The tribunal has no power to consider the validity of a compulsory purchase order. It is charged with the responsibility to determine the amount of compensation payable where land is compulsorily purchased or where compensation, in respect of depreciation in the value of the land, is claimed as a result of a public scheme. Details of the claim are registered with the tribunal (form R). A £50 fee is paid when the application is lodged.

A statement of case and reply must be made within a limited period (usually 28 days). A statement of the claimant's case is then made within 28 days. The acquiring or compensating authority then has to serve a reply within a further 28 days.

With regard to costs, the general rule is that the claimants having their land interest acquired will be awarded costs. However, this may be varied where the acquiring authority has made, a sealed offer of compensation that is not exceeded by the tribunals award, or where the claimant's conduct has unnecessarily increased the costs of the authority.

Restrictive covenants

The tribunal has the power to discharge or modify restrictive covenants within the provisions of Section 84 of the Law of Property Act 1925 (as amended). The application is registered on form LPA and a £200 fee is payable with the application. The procedures are somewhat different given that the persons possibly affected by the application may be far-reaching, ie those who have or may benefit from the covenant. More detailed information on the Lands Tribunal and procedures issued in October 2004 may be found on the Lands Tribunal website *www.landstribunal.gov.uk*. Further information will also be found on this website with regard to procedures for other actions, such as rights of light, cases under the Taxes Acts, references by consent and in respect of blight notices.

The Lands Tribunal rules 1996 (1996 No 1022)

All of the rules and procedures for actions before the tribunal are set out in this statutory instrument and a full copy can be downloaded from the same website.

Inspections

Being a specialist tribunal, it is to be expected that the member or members will wish to inspect the subject property. In my experience, this is usually taken up.

Power to order discovery (disclosure)

The tribunal (or the registrar on its behalf) is able to make suitable orders in respect of the following:

* Delivery of any document that the tribunal may require and it is in the power of the party to supply.
* Enable other parties to proceedings to inspect this document.
* Delivery to the registrar of an affidavit or make a list stating where documents or classes of documents are to be found.
* Delivery to the registrar of a statement in the form of a pleading setting out further and better particulars on the grounds on which the party intends to rely.
* To answer to interrogatories on affidavit relating to any issues between the applicant and the other party.
* Deliver to the registrar a statement of agreed facts, facts in dispute or the issues to be tried by the tribunal.
* Deliver to the registrar witness statements or proofs of evidence.

Pre-trial review

As with the case management hearings in the court, the tribunal can hold a pre-trial review following an order to that effect. The tribunal can give directions with regard to procedural matters and agreeing facts. The parties can make an application for directions at the pre-trial review (having given notice of intention to do so).

Expert witnesses

Only one expert witness on either side is allowed, unless the tribunal orders otherwise. The exceptions are hearings relating to mineral valuations or business disturbance, in which case no more than two expert witnesses on either side will be heard. However, an application for leave to call more expert witnesses may be made to the registrar.

The expert witness report must include every plan and valuation of the land or property to which proceedings relate (which shall include full particulars and computations in support of the valuations) and which it is proposed to put in evidence. Each party must also supply full particulars of any comparable properties and transactions to which the party intends to refer at the hearing in support of the statement or that no comparable properties or transactions will be referred to.

Sealed offers

Where a party wishes to protect his position on cost, there is a procedure for that party to make an unconditional offer to settle, which is enclosed in a sealed cover and sent to the registrar or delivered to the tribunal at the hearing. The sealed offer is opened by the tribunal after it has determined the case. The offer will not, of course, be disclosed until that moment.

Procedures at the hearing

The tribunal rules state that the procedure of the hearing shall be such as the tribunal may direct. In practice, the order of evidence and the procedures generally follow those employed in the High Court. In my experience, the hearing is more informal and therefore perhaps less intimidating than a High Court action. However, in a major case employing the standard or special procedures, this may not always be apparent.

In my experience, proceedings are assisted by the fact that it is a professional tribunal well versed in the type of issues in dispute and is, in consequence, very helpful in taking matters forward simply, expeditiously and fairly.

The decision of the tribunal

Most decisions are reserved at the end of the hearing. A reasoned, written decision is issued subsequently. However, in the simpler cases, the tribunal can give its decision orally. Where the decision of the tribunal is dependent on a question of law, the tribunal will state in its decision any alternative amount or value that it would have awarded if it had come to a different decision on that point of law.

Costs

The registrar will make an order as to costs. There is a right of appeal to the president within 10 days of the order.

The costs of — and incidental to — the proceedings are wholly at the discretion of the tribunal but these will normally "follow the event", as already described. The tribunal has the ability to settle the amount of costs by fixing a lump sum or directing that the costs be taxed by the registrar.

Appeals

Any appeal on a point or points of law only from the Lands Tribunal lie to the Court of Appeal but permission to appeal must be obtained.

The expert's report in a Lands Tribunal reference should, it is suggested, be prepared as recommended elsewhere in this book as in the case of any High Court action. The degree of technical explanation may be capable of some relaxation before a surveyor member, but not to the point of losing essential details of how the valuation has been prepared or omitting the valuation "trail" from comparable evidence to valuation opinion.

Valuation Tribunals

In England and Wales, the local valuation courts or tribunals have been the long-established, quasi-judicial process at first instance for determining rating appeals. The former valuation courts were replaced by the valuation and community charge tribunals on 1 May 1989. Subsequent to the passing of the Local Government Finance Act on 6 March 1992, the present day Valuation Tribunals were established, which now hear all non-domestic rating and council tax appeals in their areas.

Background

The Valuation Tribunal Service (VTS) is divided into 14 administrative units covering the whole of England and Wales under the direction of the recently appointed Valuation Tribunal Management Board, which is a non-departmental public body sponsored by the Office of the Deputy Prime Minister. Each local tribunal has an elected president, with members and elected chairmen appointed by the local rating authorities and the tribunal president. The tribunal has lay members, who may not have had relevant experience or qualifications and is assisted at each hearing by a clerk, who will advise on procedures and relevant legal aspects. Members of the Valuation Tribunals undergo appropriate training.

The administrative structure of the VTS in England is currently undergoing review and may lead to a rationalised, slimmer network of fewer tribunals and a redistribution of local hearing facilities. However, it is unlikely that the quasi-judicial processes will be changed, although the appeal process may be varied.

Constitution

The normal constitution of a Valuation Tribunal will be a chairman and two members but this can be reduced to two members with the consent of the parties. Members are expected to comply with public office "Nolan principles" in regard to selflessness, integrity, objectivity and accountability.

Members of tribunals are virtually unpaid, with only nominal expense allowances under certain circumstances.

Procedures

The majority of unresolved appeals result in a hearing before the tribunal, although it is possible, by agreement between the parties (namely the valuation officer and the ratepayer), to have the appeal disposed of by written representations. Although appeal hearings are relatively informal, they do follow the normal judicial process applicable to other courts and tribunals. Normally the appellant, whether valuation officer or ratepayer, will present his case and orally give evidence as appropriate with the opportunity for both the other party and the tribunal to ask questions. The procedure is then reversed and, finally, each party is given an opportunity to make a final summing-up statement. Evidence is not given under oath.

In practice, particularly where a ratepayer is unrepresented, the chairman with the agreement of the parties may decide to vary the procedure so that, even where the valuation officer is the respondent, he may introduce the case and the evidence first. This may be helpful to the other party and the tribunal to establish the background facts and circumstances. The tribunal will usually make every effort to assist an unrepresented party in understanding the relevant valuation facts and the law.

Decisions

The tribunal's decision is usually reserved but, in a simple, case could be given in summary at the end of the hearing after the tribunal has deliberated privately. The written decision may be transmitted as soon as reasonably practical thereafter or usually within a period of three weeks.

Appeals jurisdiction

Leaving aside council tax appeals, which is outside the scope of this book, Valuation Tribunals have the jurisdiction to hear and determine appeals concerning:

- The validity of a proposal to alter the valuation list.
- The proposal to alter the valuation list.
- The refusal to produce or permit the inspection of forms of return and particulars delivered documents (the statutory form of rent returns made to valuation officers).
- A completion notice issued by the billing authority (ie the rating authority), where a new building comes into assessment.
- A value certified by the valuation officer for the purposes of the "transitional" payment arrangements.

In addition to determining rateable values, the tribunal has the power to require the valuation officer to alter a determination or certificate value for the purposes of non-domestic rating transitional arrangements, to specify the dates where an increase in rateable value is to come into effect or where the period of the alteration is to end (for example in the case of a temporary allowance for building works).

Review of decision

On the written application of a party, the tribunal has the power to review a decision or to set it aside on the following grounds:

- A decision was wrongly made as a result of a clerical error.
- A party did not appear and can show reasonable cause why he did not do so.
- The decision is affected by a decision of the High Court or the Lands Tribunal relating to the same property.
- In the case of a completion notice, where new evidence, the existence of which could not have been ascertained by reasonably diligent inquiry or could not have been foreseen, has become available since the conclusion of the proceedings, appeals against a tribunal decision (which must be reasoned) must be made within four weeks to the Lands Tribunal.

Venues

The venue for a tribunal hearing varies according to the locality. In some rural areas, there may be a number of locations where hearings are held, usually in council offices. Some tribunals have their own courtroom facilities, such as in the case of my own Central London Valuation Tribunal.

Advocacy before tribunals

Acting for a party in a rating appeal, which proceeds to a tribunal hearing, is almost the only opportunity for the surveyor representative to present a case before a tribunal within the commercial property sector. Rating surveyors should consider carefully how the case is to be conducted and the extent to which the advocacy role is to be assumed.

In my experience as chairman of a Valuation Tribunal, valuation officers are usually quite skilled in the role, partly because of greater experience and partly because of more thorough training. Some rating surveyors are also highly skilled in presenting the client's case and circumventing the difficulties of both giving expert evidence and acting as an advocate. However, generally, standards are not good and there is scope for improvement on both sides.

If the valuer expert believes that this is an opportunity to indulge in TV drama-style courtroom antics, think again. Advocacy is a highly specialised and acquired skill and does not come easily to all surveyors. Remember that the primary duty is to the tribunal even as an advocate and to be honest, objective and helpful. Do not be over-aggressive; be courteous and explain fully complex and technical issues. Be very sure that every question to your opponent is "open" and not "closed" (ie not put in a form suggesting a particular answer). Be absolutely sure that the response desired is a virtual certainty, if not perhaps immediately in the form that you need. The best questions or succession of questions establish important information rather than attacks on personal or evidential reliability. Avoid complex questions or those that are hopefully going to adduce the right answers eventually. In my experience, they rarely do!

As highlighted earlier in this book, the expert must keep separate the roles as advocate and expert valuation witness. Expert opinion must be divorced from the presentation of the case and the persuasive elements, including relevant case-law. The rating valuer is advised to give notice to the clerk of any cases that are to be responded to in advance of the tribunal hearing (and to the other party as well).

If a valuation officer is concerned in an appeal hearing with an unrepresented ratepayer, he should try and agree facts with the ratepayer in advance. It would be beneficial to explain the rating system, the meaning of rateable value and the valuation issues that are generally involved. This will obviate the need for lengthy explanation in the tribunal, thus saving considerable time. Indeed, it may avoid the hearing altogether. In a recent case when I was chairman, the ratepayer appellant admitted that if he had received my brief explanation of the rateable value definition and basis beforehand, he would not have entered an appearance at all!

Chartered surveyors should remember that when giving evidence before the tribunal, they are bound by the *RICS Practice Statement and Guidance Note*. The necessary declarations should be made in full and the mandatory requirements adhered to (see chapter 4). In particular, be sure to separate advocacy from expert opinion and make the tribunal aware of when each role is in play.

In exceptional or major cases, the tribunal may wish to undertake an inspection following the hearing, having given notice to the parties inviting them to be present if they so desire. More information on the work of the Valuation Tribunals can be found on their website at *www.valuation-tribunals.gov.uk*.

The Future 19

Reliance and responsibility

Expert witnesses undertake a unique role in the court system and in the administration of justice, wherever their opinions are required. In the civil and criminal courts they are the one exception to the general rule that witnesses are only of fact, because of the need for the courts to receive expert testimony. As a result, they assume a high degree of responsibility. The courts, arbitrators and tribunals depend upon quality evidence on specialist and technical matters, which will often heavily influence awards and decisions. These may have substantial financial consequences in the civil arena and have major implications for the lives and liberty of human beings in the criminal courts.

The expert witness "industry" has mushroomed as a result of the growing need for technical testimony. A large part of the costs of securing justice are the fees of expert witnesses. Recent published data indicates that £130m per annum is spent by legal aid funds alone on expert witness fees. Therefore, it may be imagined that the total cost for all areas of dispute resolution must be many times this figure.

Public anxiety — Post-Cannings

The role of the expert witness and the dependence on expert testimony has attracted public and media attention. Particular concern has recently been expressed as a result of some high-profile criminal actions. Serious injustice appears to have been caused by poor-quality expert witness testimony.

I am referring to the "cot death" cases of Angela Cannings and Sally Clarke. It may be recalled that they were both imprisoned for lengthy periods for the "murder" of their babies, the convictions having relied on forensic expert evidence, which later, on appeal, was found to be wholly unreliable.

There is no doubt that these devastating miscarriages of justice could have implications for the future and may extend to the practice and conduct of expert witnesses outside the criminal code. Lord Goldsmith QC, the Attorney-General, has stated that, in his view, expert witnesses should continue to be used in criminal trials. However, they must not be partisan, the expert is there to help the jury to reach a just verdict. They must draw attention to any weaknesses in their opinions and if there are any alternative explanations. There must be better adherence to obligations on disclosure of evidence. All of this closely follows CPR obligations. As yet, there are no such parallel rules for the criminal courts.

The Legal Services Commission (LSC), which is responsible for the administration of legal aid, has already announced its belief that, in the long term, all experts will need to be accredited if they are to be paid from public funds.

LSC consultation

In November 2004, the LSC produced a consultation paper that not only produces guidelines in respect of civil proceedings, but also introduces, for the first time, the concept of certification, but only for forensic experts who regularly provide expert evidence. They should be quality assured, ie "accredited".

Statutory or self-regulation?

Some expert witness bodies have already announced schemes for voluntary registration or accreditation. Over time, it must be expected that some statutory interference may evolve. It is, indeed, unfortunate that many professional institutions have not established the RICS-style codes of practice to be found in the *Practice Statement and Guidance Note*. It must be anticipated that greater regulation by professional bodies will take place but, possibly, this will not be sufficient on its own.

We live in an age of media-driven debate on issues of public concern and perception, leading often to "correct" solutions rather than what might be best practice. Political expediency often takes precedence. It is hoped that self-regulation and greater adherence to codes of practice, including CPR, will suffice but, in my view, it will only take a major incident similar to Cannings in the civil arena for some form of statutory intervention to become virtually certain. This would be a backward step, in my opinion.

There is a significant further dimension to this debate. Many commentators have considered that part of the problem resulting in the Cannings misadventure is the whole system of administering justice using adversarial court procedures. The latter may tend to encourage the abuse of legal process, such as non-disclosure of important evidence or facts. It will have been observed throughout this book that we appear to be moving closer to the inquisitorial and away from the adversarial but, fundamentally, the latter still prevails.

There is substantial cultural resistance within the legal profession to any substantial move away from the present system. This is hardly surprising since it is likely that any fundamental change could wreak havoc with the practices of litigation solicitors and barristers.

Expert witnesses would certainly be caught up in any changes, moving further away from expert testimony to expert advice.

Whatever the eventual outcome of this debate, it is believed that one thing is certain: there will be fewer cases outside the criminal courts proceeding to a trial. Partly as a result of the expert witness debate and the perceived substantial costs and risks of litigation, parties will be encouraged to try alternative dispute resolution by mediation or otherwise. Thus, the expert's role as adviser will be further re-enforced. There may be less concern, however, in taking cases to arbitration or tribunals.

I have no doubt that whatever the future holds, better-quality valuation evidence and advice from surveyors will be required, the key to which must be partly better education and training.

Fortunately, the aspiring expert witness surveyor will have no difficulty in obtaining such training and keeping up to date with best practice, case-law and techniques. A number of bodies offer expert witness training, including report writing and courtroom skills. In addition, expert witness professional institutions and the RICS regularly run courses and seminars. The latter has established an Expert Witness Registration Scheme. Also, the major law practices offer a wide variety of seminars on related subjects. In this respect, the surveyor expert is truly spoilt for choice.

Appendix 1

CPR Part 35 and Practice Direction (PD 35)

Civil Procedure Rules
Part 35
Experts and Assessors

Contents of this part

Duty to restrict expert evidence

35.1   Expert evidence shall be restricted to that which is reasonably required to resolve the proceedings.

35.2   A reference to an 'expert' in this Part is a reference to an expert who has been instructed to give or prepare evidence for the purpose of court proceedings.

Experts — overriding duty to the court

35.3 (1)   It is the duty of an expert to help the court on the matters within his expertise.

(2)   This duty overrides any obligation to the person from whom he has received instructions or by whom he is paid.

Court's power to restrict expert evidence

35.4 (1)   No party may call an expert or put in evidence an expert's report without the court's permission.

(2)   When a party applies for permission under this rule he must identify—

(a) the field in which he wishes to rely on expert evidence; and

(b) where practicable the expert in that field on whose evidence he wishes to rely.

(3)   If permission is granted under this rule it shall be in relation only to the expert named or the field identified under paragraph (2).

(4)   The court may limit the amount of the expert's fees and expenses that the party who wishes to rely on the expert may recover from any other party.

General requirement for expert evidence to be given in a written report

35.5 (1)   Expert evidence is to be given in a written report unless the court directs otherwise.

(2)   If a claim is on the fast track, the court will not direct an expert to attend a hearing unless it is necessary to do so in the interests of justice.

Written questions to experts

35.6 (1)   A party may put to —

(a)   an expert instructed by another party; or

(b)   a single joint expert appointed under rule 35.7,

written questions about his report.

(2)   Written questions under paragraph (1) —

(a)   may be put once only;

(b)   must be put within 28 days of service of the expert's report; and

(c)   must be for the purpose only of clarification of the report,

unless in any case —

(i)   the court gives permission; or

(ii)   the other party agrees.

(3)   An expert's answers to questions put in accordance with paragraph (1) shall be treated as part of the expert's report.

(4)   Where —

(a)   a party has put a written question to an expert instructed by another party in accordance with this rule; and

(b)   the expert does not answer that question,

the court may make one or both of the following orders in relation to the party who instructed the expert —

(i)   that the party may not rely on the evidence of that expert; or

(ii)   that the party may not recover the fees and expenses of that expert from any other party.

Court's power to direct that evidence is to be given by a single joint expert

35.7 (1)   Where two or more parties wish to submit expert evidence on a particular issue, the court may direct that the evidence on that issue is given by one expert only.

(2)   The parties wishing to submit the expert evidence are called 'the instructing parties'.

(3)   Where the instructing parties cannot agree who should be the expert, the court may —

(a)   select the expert from a list prepared or identified by the instructing parties; or

(b)   direct that the expert be selected in such other manner as the court may direct.

Instructions to a single joint expert

35.8 (1)   Where the court gives a direction under rule 35.7 for a single joint expert to be used, each instructing party may give instructions to the expert.

(2)   When an instructing party gives instructions to the expert he must, at the same time, send a copy of the instructions to the other instructing parties.

(3)   The court may give directions about —

(a)   the payment of the expert's fees and expenses; and

(b)   any inspection, examination or experiments which the expert wishes to carry out.

(4)   The court may, before an expert is instructed —

(a)   limit the amount that can be paid by way of fees and expenses to the expert; and

(b)   direct that the instructing parties pay that amount into court.

(5)   Unless the court otherwise directs, the instructing parties are jointly and severally liable (GL) for the payment of the expert's fees and expenses.

Power of court to direct a party to provide information

35.9   Where a party has access to information which is not reasonably available to the other party, the court may direct the party who has access to the information to —

  (a) prepare and file and document recording the information; and

  (b) serve a copy of that document on the other party.

Contents of report

35.10 (1)   An expert's report must comply with the requirements set out in the relevant practice direction.

  (2)   At the end of an expert's report there must be a statement that —

    (a) the expert understands his duty to the court; and

    (b) he has complied with that duty.

  (3)   The expert's report must state the substance of all material instructions, whether written or oral, on the basis of which the report was written.

  (4)   The instructions referred to in paragraph (3) shall not be privileged (GL) against disclosure but the court will not, in relation to those instructions —

    (a) order disclosure of any specific document; or

    (b) permit any questioning in court, other than by the party who instructed the expert,

unless it is satisfied that there are reasonable grounds to consider the statement of instructions given under paragraph (3) to be inaccurate or incomplete.

Use by one party of expert's report disclosed by another

35.11   Where a party has disclosed an expert's report, any party may use that expert's report as evidence at the trial.

Discussions between experts

35.12 (1)   The court may, at any stage, direct a discussion between experts for the purpose of requiring the experts to —

    (a) identify and discuss the expert issues in the proceedings; and

      (b) where possible, reach an agreed opinion on those issues.

(2)   The court may specify the issues which the experts must discuss.

(3)   The court may direct that following a discussion between the experts they must prepare a statement for the court showing —

      (a) those issues on which they agree; and

      (b) those issues on which they disagree and a summary of their reasons for disagreeing.

(4)   The content of the discussion between the experts shall not be referred to at the trial unless the parties agree.

(5)   Where experts reach agreement on an issue during their discussions, the agreement shall not bind the parties unless the parties expressly agree to be bound by the agreement.

Consequence of failure to disclose expert's report

35.13   A party who fails to disclose an expert's report may not use the report at the trial or call the expert to give evidence orally unless the court gives permission.

Expert's right to ask court for directions

13.14 (1)   An expert may file a written request for directions to assist him in carrying out his function as an expert

(2)   An expert must, unless the court orders otherwise, provide a copy of any proposed request for directions under paragraph (1) —

      (a) to the party instructing him, at least 7 days before he files the request; and

      (b) to all other parties, at least 4 days before he files it.

(3)   The court, when it gives directions, may also direct that a party be served with a copy of the directions.

Assessors

13.15 (1)   This rule applies where the court appoints one or more persons (an 'assessor') under section 70 of the Supreme

Court Act 1981 [1] or section 63 of the County Courts Act 1984 [2].

(2)   The assessor shall assist the court in dealing with a matter in which the assessor has skill and experience.

(3)   An assessor shall take such part in the proceedings as the court may direct and in particular the court may —

    (a)   direct the assessor to prepare a report for the court on any matter at issue in the proceedings; and

    (b)   direct the assessor to attend the whole or any part of the trial to advise the court on any such matter.

(4)   If the assessor prepares a report for the court before the trial has begun —

    (a)   the court will send a copy to each of the parties; and

    (b)   the parties may use it at trial.

(5)   The remuneration to be paid to the assessor for his services shall be determined by the court and shall form part of the costs of the proceedings.

(6)   The court may order any party to deposit in the court office a specified sum in respect of the assessor's fees and, where it does so, the assessor will not be asked to at until the sum has been deposited.

(7)   Paragraphs (5) and (6) do not apply where the remuneration of the assessor is to be paid out of money provided by Parliament.

Footnotes

1   1981 c.54.
2   1984 c.28. Section 63 was amended by S.I. 1988/2940.

Practice Direction — Experts and Assessors
This practice direction supplements CPR Part 35

Contents of this Practice Direction

Part 35 is intended to limit the use of oral expert evidence to that which is reasonably required. In addition, where possible, matters requiring expert evidence should be dealt with by a single expert. Permission of the court is always required either to call an expert or to put an expert's report in evidence.

Expert evidence — general requirements

1.1 It is the duty of an expert to help the court on matters within his own expertise: rule 35.3(1). This duty is paramount and overrides any obligation to the person from whom the expert has received instructions or by whom he is paid: rule 35.3(2).

1.2 Expert evidence should be the independent product of the expert uninfluenced by the pressures of litigation.

1.3 An expert should assist the court by providing objective, unbiased opinion on matters within his expertise, and should not assume the role of an advocate.

1.4 An expert should consider all material facts, including those which might detract from his opinion.

1.5 An expert should make it clear:
   (a) when a question or issue falls outside his expertise; and
   (b) when he is not able to reach a definite opinion, for example because he has insufficient information.

1.6 If, after producing a report, an expert changes his view on any material matter, such change of view should be communicated to all the parties without delay, and when appropriate to the court.

Form and content of expert's reports

2.1 An expert's report should be addressed to the court and not to the party from whom the expert has received his instructions.

2.2 An expert's report must:
  (1) give details of the expert's qualifications;
  (2) give details of any literature or other material which the expert has relied on in making the report;
  (3) contain a statement setting out the substance of all facts and instructions given to the expert which are material to the opinions expressed in the report or upon which those opinions are based
  (4) make clear which of the facts stated in the report are within the expert's own knowledge;
  (5) say who carried out any examination, measurement, test or experiment which the expert has used for the report, give the qualification of that person, and say whether or not the test or experiment has been carried out under the expert's supervision;
  (6) where there is a range of opinion on the matters dealt with in the report—
    (a) summarise the range of opinion, and
    (b) give reasons for his own opinion;
  (7) contain a summary of the conclusions reached;
  (8) if the expert is not able to give his opinion without qualification, state the qualification; and
  (9) contain a statement that the expert understands his duty to the court, and has complied and will continue to comply with that duty.

2.3 An expert's report must be verified by a statement of truth as well as containing the statements required in paragraph 2.2(8) and (9) above.

2.4 The form of the statement of truth is as follows:
"I confirm that insofar as the facts stated in my report are within my own knowledge I have made clear which they are and I believe them to be true, and that the opinions I have expressed represent my true and complete professional opinion."

2.5 Attention is drawn to rule 32.14 which sets out the consequences of verifying a document containing a false statement without an honest belief in its truth.
(For information about statements of truth see Part 22 and the practice direction which supplements it.)

Information

3   Under Rule 35.9 the court may direct a party with access to information which is not reasonably available to another party to serve on that other party a document which records the information. The document served must include sufficient details of all the facts, tests, experiments ad assumptions which underlie any part of the information to enable the party on whom it is served to make, or to obtain, a proper interpretation of the information and an assessment of its significance.

Instructions

4   The instructions referred to in paragraph 2.2(3) will not be protected by privilege (see rule 35.10(4)). But cross-examination of the expert on the contents of his instructions will not be allowed unless the court permits it (or unless the party who gave the instructions consents to it). Before it gives permission the court must be satisfied that there are reasonable grounds to consider that the statement in the report of the substance of the instructions is inaccurate or incomplete. If the court is so satisfied, it will allow the cross-examination where it appears to be in the interests of justice to do so.

Questions to experts

5.1   Questions asked for the purpose of clarifying the expert's report (see rule 35.6) should be put, in writing, to the expert not later than 28 days after receipt of the expert's report (see paragraphs 1.2 to 1.5 above as to verification).

5.2   Where a party sends a written question or questions direct to an expert, a copy of the questions should, at the same time, be sent to the other party or parties.

5.3   The party or parties instructing the expert must pay any fees charged by that expert for answering questions put under rule 35.6. This does not affect any decision of the court as to the party who is ultimately to bear the expert's costs.

Single expert

6   Where the court has directed that the evidence on a particular
     issue is to be given by one expert only (rule 35.7) but there are a
     number of disciplines relevant to that issue, a leading expert in the
     dominant discipline should be identified as the single expert. He
     should prepare the general part of the report and be responsible
     for annexing or incorporating the contents of any reports from
     experts in other disciplines.

Assessors

7.1   An assessor may be appointed to assist the court under rule 35.15.
      Not less than 21 days before making any such appointment, the
      court will notify each party in writing of the name of the proposed
      assessor, of the matter in respect of which the assistance of the
      assessor will be sought and of the qualifications of the assessor to
      give that assistance.

7.2   Where any person has been proposed for appointment as an
      assessor, objection to him, either personally or in respect of his
      qualification, may be taken by any party.

7.3   Any such objection must be made in writing and field with the
      court within 7 days of receipt of the notification referred to in
      paragraph 6.1 and will be taken into account by the court in
      deciding whether or not to make the appointment (section 63(5) of
      the County Courts Act 1984).

7.4   Copies of any report prepared by the assessor will be sent to each
      of the parties but the assessor will not give oral evidence or be
      open to cross-examination or questioning.

Appendix 2

Protocol for the Instruction of Experts to give Evidence in Civil Claims

Contents

Protocol for the Instruction of Experts to give evidence in civil claims

1. Introduction

Expert witnesses perform a vital role in civil litigation. It is essential that both those who instruct experts and experts themselves are given clear guidance as to what they are expected to do in civil proceedings. The purpose of this Protocol is to provide such guidance. It has been drafted by the Civil Justice Council and reflects the rules and practice directions current [in June 2005], replacing the Code of Guidance on Expert Evidence. The authors of the Protocol wish to acknowledge the valuable assistance they obtained by drawing on earlier documents produced by the Academy of Experts and the Expert Witness Institute, as well as suggestions made by the Clinical Dispute Forum. The Protocol has been approved by the Master of the Rolls.

2. Aims of Protocol

2.1 This Protocol offers guidance to experts and to those instructing them in the interpretation of and compliance with Part 35 of the Civil Procedure Rules (CPR 35) and its associated Practice Direction (PD 35) and to further the objectives of the Civil Procedure Rules in general. It is intended to assist in the interpretation of those provisions in the interests of good practice but it does not replace them. It sets out standards for the use of experts and the conduct of experts and those who instruct them. The existence of this Protocol does not remove the need for experts and those who instruct them to be familiar with CPR35 and PD35.

2.2 Experts and those who instruct them should also bear in mind para 1.4 of the Practice Decision on Protocols which contains the following objectives, namely to:

(a) encourage the exchange of early and full information about the expert issues involved in a prospective legal claim;

(b) enable the parties to avoid or reduce the scope of the litigation by agreeing the whole or part of an expert issue before commencement of proceedings; and

(c) support the efficient management of proceedings where litigation cannot be avoided.

3. Application

3.1 This Protocol applies to any steps taken for the purpose of civil proceedings by experts or those who instruct them on or after 5th September 2004.

3.2 It applies to all experts who are, or who may be, governed by CPR Part 35 and to those who instruct them. Experts are governed by Part 35 if they are or have been instructed to give or prepare evidence for the purpose of civil proceedings in a court in England and Wales (CPR 35.2).

3.3 Experts, and those instructing them, should be aware that some cases may be "specialist proceedings" (CPR 49) where there are modifications to the Civil Procedure Rules. Proceedings may also be governed by other Protocols. Further, some courts have published their own Guides which supplement the Civil Procedure Rules for proceedings in those courts. They contain provisions affecting expert evidence. Expert witnesses and those instructing them should be familiar with them when they are relevant.

3.4 Courts may take into account any failure to comply with this Protocol when making orders in relation to costs, interest, time limits, the stay of proceedings and whether to order a party to pay a sum of money into court.

Limitation

3.5 If, as a result of complying with any part of this Protocol, claims would or might be time barred under any provision in the Limitation Act 1980, or any other legislation that imposes a time limit for the bringing an action, claimants may commence proceedings without complying with this Protocol. In such circumstances claimants who commence proceedings without complying with all, or any part, of this Protocol must apply, giving notice to all other parties, to the court for directions as to the timetable and form of procedure to be adopted, at the same time as they request the court to issue proceedings. The court may consider whether to order a stay of the whole or part of the proceedings pending compliance with this Protocol and may make orders in relation to costs.

4. Duties of experts

4.1 Experts always owe a duty to exercise reasonable skill and care to those instructing them, and to comply with any relevant professional code of ethics. However when they are instructed to give or prepare evidence for the purpose of civil proceedings in England and Wales they have an overriding duty to help the court on matters within their expertise (CPR 35.3). This duty overrides any obligation to the person instructing or paying them. Experts must not serve the exclusive interest of those who retain them.

4.2 Experts should be aware of the overriding objective that courts deal with cases justly. This includes dealing with cases proportionately, expeditiously and fairly (CPR 1.1). Experts are under an obligation to assist the court so as to enable them to deal with cases in accordance with the overriding objective. However the overriding objective does not impose on experts any duty to act as mediators between the parties or require them to trespass on the role of the court in deciding facts.

4.3 Experts should provide opinions which are independent, regardless of the pressures of litigation. In this context, a useful test of "independence" is that the expert would express the same opinion if given the same instructions by an opposing party. Experts should not take it upon themselves to promote the point of view of the party instructing them or engage in the role of advocates.

4.4 Experts should confine their opinions to matters which are material to the disputes between the parties and provide opinions only in relation to matters which lie within their expertise. Experts should indicate without delay where particular questions or issues fall outside their expertise.

4.5 Experts should take into account all material facts before them at the time that they give their opinion. Their reports should set out those facts and any literature or any other material on which they have relied in forming their opinions. They should indicate if an opinion is provisional, or qualified, or where they consider that further information is required or if, for any other reason, they are not satisfied that an opinion can be expressed finally and without qualification.

4.6 Experts should inform those instructing them without delay of any change in their opinions on any material matter and the reason for it.

4.7 Experts should be aware that any failure by them to comply with the Civil Procedure Rules or court orders or any excessive delay for which they are responsible may result in the parties who instructed them being penalised in costs and even, in extreme cases, being debarred from placing the experts' evidence before the court. In *Phillips* v *Symes*[1] Peter Smith J held that courts may also make orders for costs (under section 51 of the Supreme Court Act 1981) directly against expert witnesses who by their evidence cause significant expense to be incurred, and do so in flagrant and reckless disregard of their duties to the Court.

5. Conduct of Experts instructed only to advise

5.1 Part 35 only applies where experts are instructed to give opinions which are relied on for the purposes of court proceedings. Advice

---

[1]   *Phillips* v *Symes* [2004] EWHC 2330 (Ch).

which the parties do not intend to adduce in litigation is likely to be confidential; the Protocol does not apply in these circumstances.[2,3]

5.2 The same applies where, after the commencement of proceedings, experts are instructed only to advise (e.g. to comment upon a single joint expert's report) and not to give or prepare evidence for use in the proceedings.

5.3 However this Protocol does apply if experts who were formerly instructed only to advise are later instructed to give or prepare evidence for the purpose of civil proceedings.

6. The Need for Experts

6.1 Those intending to instruct experts to give or prepare evidence for the purpose of civil proceedings should consider whether expert evidence is appropriate, taking account of the principles set out in CPR Parts 1 and 35, and in particular whether:

(a) it is relevant to a matter which is in dispute between the parties;
(b) it is reasonably required to resolve the proceedings (CPR 35.1);
(c) the expert has expertise relevant to the issue on which an opinion is sought;
(d) the expert has the experience, expertise and training appropriate to the value, complexity and importance of the case; and whether
(e) these objects can be achieved by the appointment of a single joint expert (see section 17 below).

6.2 Although the court's permission is not generally required to instruct an expert, the court's permission is required before experts can be called to give evidence or their evidence can be put in (CPR 35.4).

7. The appointment of experts

7.1 Before experts are formally instructed or the court's permission to appoint named experts is sought, the following should be established:

(a) that they have the appropriate expertise and experience;
(b) that they are familiar with the general duties of an expert;
(c) that they can produce a report, deal with questions and have

---

[2] *Carlson* v *Townsend* [2001] 1 WLR 2415.
[3] *Jackson* v *Marley Davenport* [2004] 1 WLR 2926.

discussions with other experts within a reasonable time and at a cost proportionate to the matters in issue;

(d) a description of the work required;

(e) whether they are available to attend the trial, if attendance is required; and

(f) there is no potential conflict of interest.

7.2 Terms of appointment should be agreed at the outset and should normally include:

(a) the capacity in which the expert is to be appointed (e.g. party appointed expert, single joint expert or expert advisor);

(b) the services required of the expert (e.g. provision of expert's report, answering questions in writing, attendance at meetings and attendance at court);

(c) time for delivery of the report;

(d) the basis of the expert's charges (either daily or hourly rates and an estimate of the time likely to be required, or a total fee for the services);

(e) travelling expenses and disbursements;

(f) cancellation charges;

(g) any fees for attending court;

(h) time for making the payment; and

(i) whether fees are to be paid by a third party

(j) if a party is publicly funded, whether or not the expert's charges will be subject to assessment by a costs officer.

7.3 As to the appointment of single joint experts, see section 17 below.

7.4 When necessary, arrangements should be made for dealing with questions to experts and discussions between experts, including any directions given by the court, and provision should be made for the cost of this work.

7.5 Experts should be informed regularly about deadlines for all matters concerning them. Those instructing experts should promptly send them copies of all court orders and directions which may affect the preparation of their reports or any other matters concerning their obligations.

Conditional and Contingency Fees

7.6 Payments contingent upon the nature of the expert evidence given in legal proceedings, or upon the outcome of a case, must not be offered or accepted. To do so would contravene experts' overriding duty to the court and compromise their duty of independence.

7.7 Agreement to delay payment of experts' fees until after the conclusion of cases is permissible as long as the amount of the fee does not depend on the outcome of the case.

8. Instructions

8.1 Those instructing experts should ensure that they give clear instructions, including the following:

(a) basic information, such as names, addresses, telephone numbers, dates of birth and dates of incidents;
(b) the nature and extent of the expertise which is called for;
(c) the purpose of requesting the advice or report, a description of the matter(s) to be investigated, the principal known issues and the identity of all parties;
(d) the statement(s) of case (if any), those documents which form part of standard disclosure and witness statements which are relevant to the advice or report;
(e) where proceedings have not been started, whether proceedings are being contemplated and, if so, whether the expert is asked only for advice;
(f) an outline programme, consistent with good case management and the expert's availability, for the completion and delivery of each stage of the expert's work; and
(g) where proceedings have been started, the dates of any hearings (including any Case Management Conferences and/or Pre-Trial Reviews), the name of the court, the claim number and the track to which the claim has been allocated.

8.2 Experts who do not receive clear instructions should request clarification and may indicate that they are not prepared to act unless and until such clear instructions are received.

8.3 As to the instruction of single joint experts, see section 17 below.

9. Experts' Acceptance of Instructions

9.1 Experts should confirm without delay whether or not they accept instructions. They should also inform those instructing them (whether on initial instruction or at any later stage) without delay if:

(a)   instructions are not acceptable because, for example, they require work that falls outside their expertise, impose unrealistic deadlines, or are insufficiently clear;

(b)   they consider that instructions are or have become insufficient to complete the work;

(c)   they become aware that they may not be able to fulfil any of the terms of appointment;

(d)   the instructions and/or work have, for any reason, placed them in conflict with their duties as an expert; or

(e)   they are not satisfied that they can comply with any orders that have been made.

9.2 Experts must neither express an opinion outside the scope of their field of expertise, nor accept any instructions to do so.

10. Withdrawal

10.1 Where experts' instructions remain incompatible with their duties, whether through incompleteness, a conflict between their duty to the court and their instructions, or for any other substantial and significant reason, they may consider withdrawing from the case. However, experts should not withdraw without first discussing the position fully with those who instruct them and considering carefully whether it would be more appropriate to make a written request for directions from the court. If experts do withdraw, they must give formal written notice to those instructing them.

11. Experts' Right to ask Court for Directions

11.1 Experts may request directions from the court to assist them in carrying out their functions as experts. Experts should normally discuss such matters with those who instruct them before making any such request. Unless the court otherwise orders, any proposed request for directions should be copied to the party instructing the expert at least seven days before filing any request to the court, and to all other parties at least four days before filing it. (CPR 35.14).

11.2 Requests to the court for directions should be made by letter, containing

(a)   the title of the claim;

(b)   the claim number of the case;

(c)   the name of the expert;

(d)   full details of why directions are sought; and

(e)   copies of any relevant documentation.

12. Power of the Court to Direct a Party to Provide Information

12.1 If experts consider that those instructing them have not provided information which they require, they may, after discussion with those instructing them and giving notice, write to the court to seek directions. (CPR 35.14).

12.2 Experts and those who instruct them should also be aware of CPR 35.9. This provides that where one party has access to information which is not readily available to the other party, the court may direct the party who has access to the information to prepare, file and copy to the other party a document recording the information. If experts require such information which has not been disclosed, they should discuss the position with those instructing them without delay, so that a request for the information can be made, and, if not forthcoming, an application can be made to the court. Unless a document appears to be essential, experts should assess the cost and time involved in the production of a document and whether its provision would be proportionate in the context of the case.

13. Contents of Experts' Reports

13.1 The content and extent of experts' reports should be governed by the scope of their instructions and general obligations, the contents of CPR 35 and PD35 and their overriding duty to the court.

13.2 In preparing reports, experts should maintain professional objectivity and impartiality at all times.

13.3 PD 35, para 2 provides that experts' reports should be addressed to the court and gives detailed directions about the form and content of such reports. All experts and those who instruct them should ensure that they are familiar with these requirements.

13.4 Model forms of Experts' Reports are available from bodies such as the Academy of Experts or the Expert Witness Institute.

13.5 Experts' reports must contain statements that they understand their duty to the court and have complied and will continue to comply with that duty (PD35 para 2.2(9)). They must also be verified by a statement of truth. The form of the statement of truth is as follows:

> I confirm that insofar as the facts stated in my report are within my own knowledge I have made clear which they are and I believe them to be true, and that the opinions I have expressed represent my true and complete professional opinion.

This wording is mandatory and must not be modified.

Qualifications

13.6 The details of experts' qualifications to be given in reports should be commensurate with the nature and complexity of the case. It may be sufficient merely to state academic and professional qualifications. However, where highly specialised expertise is called for, experts should include the detail of particular training and/or experience that qualifies them to provide that highly specialised evidence.

Tests

13.7 Where tests of a scientific or technical nature have been carried out, experts should state:

(a) the methodology used; and
(b) by whom the tests were undertaken and under whose supervision, summarising their respective qualifications and experience.

Reliance on the work of others

13.8 Where experts rely in their reports on literature or other material and cite the opinions of others without having verified them, they must give details of those opinions relied on. It is likely to assist the court if the qualifications of the originator(s) are also stated.

Facts

13.9 When addressing questions of fact and opinion, experts should keep the two separate and discrete.

13.10 Experts must state those facts (whether assumed or otherwise) upon which their opinions are based. They must distinguish clearly between those facts which experts know to be true and those facts which they assume.

13.11 Where there are material facts in dispute experts should express separate opinions on each hypothesis put forward. They should not express a view in favour of one or other disputed version of the facts unless, as a result of particular expertise and experience, they consider one set of facts as being improbable or less probable, in which case they may express that view, and should give reasons for holding it.

Range of opinion

13.12 If the mandatory summary of the range of opinion is based on published sources, experts should explain those sources and, where appropriate, state the qualifications of the originator(s) of the opinions from which they differ, particularly if such opinions represent a well-established school of thought.

13.13 Where there is no available source for the range of opinion, experts may need to express opinions on what they believe to be the range which other experts would arrive at if asked. In those circumstances, experts should make it clear that the range that they summarise is based on their own judgement and explain the basis of that judgement.

Conclusions

13.14 A summary of conclusions is mandatory. The summary should be at the end of the report after all the reasoning. There may be cases, however, where the benefit to the court is heightened by placing a short summary at the beginning of the report whilst giving the full conclusions at the end. For example, it can assist with the comprehension of the analysis and with the absorption of the detailed facts if the court is told at the outset of the direction in which the report's logic will flow in cases involving highly complex matters which fall outside the general knowledge of the court.

Basis of report: material instructions

13.15 The mandatory statement of the substance of all material instructions should not be incomplete or otherwise tend to mislead. The imperative is transparency. The term "instructions" includes all material which solicitors place in front of experts in order to gain advice. The omission from the statement of "off-the-record" oral instructions is not permitted. Courts may allow cross-examination about the instructions if there are reasonable grounds to consider that the statement may be inaccurate or incomplete.

14. After receipt of experts' reports

14.1 Following the receipt of experts' reports, those instructing them should advise the experts as soon as reasonably practicable whether, and if so when, the report will be disclosed to other parties; and, if so disclosed, the date of actual disclosure.

14.2 If experts' reports are to be relied upon, and if experts are to give oral evidence, those instructing them should give the experts the opportunity to consider and comment upon other reports within their area of expertise and which deal with relevant issues at the earliest opportunity.

14.3 Those instructing experts should keep experts informed of the progress of cases, including amendments to statements of case relevant to experts' opinion.

14.4 If those instructing experts become aware of material changes in circumstances or that relevant information within their control was not previously provided to experts, they should without delay instruct experts to review, and if necessary, update the contents of their reports.

15. Amendments of reports

15.1 It may become necessary for experts to amend their reports:

(a)   as a result of an exchange of questions and answers;
(b)   following agreements reached at meetings between experts; or
(c)   where further evidence or documentation is disclosed.

15.2 Experts should not be asked to, and should not, amend, expand or alter any parts of reports in a manner which distorts their true opinion, but may be invited to amend or expand reports to ensure accuracy, internal consistency, completeness and relevance to the issues and clarity. Although experts should generally follow the recommendations of solicitors with regard to the form of reports, they should form their own independent views as to the opinions and contents expressed in their reports and exclude any suggestions which do not accord with their views.

15.3 Where experts change their opinion following a meeting of experts, a simple signed and dated addendum or memorandum to that effect is generally sufficient. In some cases, however, the benefit to the court of having an amended report may justify the cost of making the amendment.

15.4 Where experts significantly alter their opinion, as a result of new evidence or because evidence on which they relied has become unreliable, or for any other reason, they should amend their reports to reflect that fact. Amended reports should include reasons for amendments. In such circumstances, those instructing experts should inform other parties as soon as possible of any change of opinion.

15.5 When experts intend to amend their reports, they should inform those instructing them without delay and give reasons. They should provide the amended version (or an addendum or memorandum) clearly marked as such as quickly as possible.

16. Written Questions to Experts

16.1 The procedure for putting written questions to experts (CPR 35.6) is intended to facilitate the clarification of opinions and issues after experts' reports have been served. Experts have a duty to provide answers to questions properly put. Where they fail to do so, the court may impose sanctions against the party instructing the expert, and, if there is continued non-compliance, debar a party from relying on the report. Experts should copy their answers to those instructing them.

16.2 Experts' answers to questions automatically become part of their reports. They are covered by the statement of truth and form part of the expert evidence.

16.3 Where experts believe that questions put are not properly directed to the clarification of the report, or are disproportionate, or have been asked out of time, they should discuss the questions with those instructing them and, if appropriate, those asking the questions. Attempts should be made to resolve such problems without the need for an application to the court for directions.

Written requests for directions in relation to questions

16.4 If those instructing experts do not apply to the court in respect of questions, but experts still believe that questions are improper or out of time, experts may file written requests with the court for directions to assist in carrying out their functions as experts (CPR 35.14). See section 11 above.

17. Single Joint Experts

17.1 CPR 35 and PD35 deal extensively with the instruction and use of joint experts by the parties and the powers of the court to order their use (see CPR 35.7 and 35.8, PD35, para 5).

17.2 The Civil Procedure Rules encourage the use of joint experts. Wherever possible a joint report should be obtained. Consideration should therefore be given by all parties to the appointment of single joint experts in all cases where a court might direct such an appointment. Single joint experts are the norm in cases allocated to the small claims track and the fast track.

17.3 Where, in the early stages of a dispute, examinations, investigations, tests, site inspections, experiments, preparation of photographs, plans or other similar preliminary expert tasks are necessary, consideration should be given to the instruction of a single joint expert, especially where such matters are not, at that stage, expected to be contentious as between the parties. The objective of such an appointment should be to agree or to narrow issues.

17.5 Experts who have previously advised a party (whether in the same case or otherwise) should only be proposed as single joint experts if other parties are given all relevant information about the previous involvement.

17.6 The appointment of a single joint expert does not prevent parties from instructing their own experts to advise (but the costs of such expert advisers may not be recoverable in the case).

Joint instructions

17.7 The parties should try to agree joint instructions to single joint experts, but, in default of agreement, each party may give instructions. In particular, all parties should try to agree what documents should be included with instructions and what assumptions single joint experts should make.

17.8 Where the parties fail to agree joint instructions, they should try to agree where the areas of disagreement lie and their instructions should make this clear. If separate instructions are given, they should be copied at the same time to the other instructing parties.

17.9 Where experts are instructed by two or more parties, the terms of appointment should, unless the court has directed otherwise, or the parties have agreed otherwise, include:

(a) a statement that all the instructing parties are jointly and severally liable to pay the experts' fees and, accordingly, that experts' invoices should be sent simultaneously to all instructing parties or their solicitors (as appropriate); and

(b) a statement as to whether any order has been made limiting the amount of experts' fees and expenses (CPR 35.8(4)(a)).

17.10 Where instructions have not been received by the expert from one or more of the instructing parties the expert should give notice (normally at least 7 days) of a deadline to all instructing parties for the receipt by the expert of such instructions. Unless the instructions are received within the deadline the expert may begin work. In the event that instructions are received after the deadline but before the signing off of the report the expert should consider whether it is practicable to comply with those instructions without adversely affecting the timetable set for delivery of the report and in such a manner as to comply with the proportionality principle. An expert who decides to issue a report without taking into account instructions received after the deadline should inform the parties who may apply to the court for directions. In either event the report must show clearly that the expert did not receive instructions within the deadline, or, as the case may be, at all.

Conduct of the single joint expert

17.11 Single joint experts should keep all instructing parties informed of any material steps that they may be taking by, for example, copying all correspondence to those instructing them.

17.12 Single joint experts are Part 35 experts and so have an overriding duty to the court. They are the parties' appointed experts and therefore owe an equal duty to all parties. They should maintain independence, impartiality and transparency at all times.

17.13 Single joint experts should not attend any meeting or conference which is not a joint one, unless all the parties have agreed in writing or the court has directed that such a meeting may be held and who is to pay the experts' fees for the meeting.

17.14 Single joint experts may request directions from the court — see Section 11 above.

17.15 Single joint experts should serve their reports simultaneously on all instructing parties. They should provide a single report even though they may have received instructions which contain areas of conflicting fact or allegation. If conflicting instructions lead to different opinions (for example, because the instructions require experts to make different assumptions of fact), reports may need to contain more than one set of opinions on any issue. It is for the court to determine the facts.

Cross-examination

17.16 Single joint experts do not normally give oral evidence at trial but if they do, all parties may cross-examine them. In general written questions (CPR 35.6) should be put to single joint experts before requests are made for them to attend court for the purpose of cross-examination.[4]

18. Discussions between Experts

18.1 The court has powers to direct discussions between experts for the purposes set out in the Rules (CPR 35.12). Parties may also agree that discussions take place between their experts.

18.2 Where single joint experts have been instructed but parties have,

---

[4]   *Peet* v *Mid Kent Area Healthcare NHS Trust* [2002] 1 WLR 210.

with the permission of the court, instructed their own additional Part 35 experts, there may, if the court so orders or the parties agree, be discussions between the single joint experts and the additional Part 35 experts. Such discussions should be confined to those matters within the remit of the additional Part 35 experts or as ordered by the court.

18.3 The purpose of discussions between experts should be, wherever possible, to:

(a) identify and discuss the expert issues in the proceedings;
(b) reach agreed opinions on those issues, and, if that is not possible, to narrow the issues in the case;
(c) identify those issues on which they agree and disagree and summarise their reasons for disagreement on any issue; and
(d) identify what action, if any, may be taken to resolve any of the outstanding issues between the parties.

Arrangements for discussions between experts

18.4 Arrangements for discussions between experts should be proportionate to the value of cases. In small claims and fast-track cases there should not normally be meetings between experts. Where discussion is justified in such cases, telephone discussion or an exchange of letters should, in the interests of proportionality, usually suffice. In multi-track cases, discussion may be face to face, but the practicalities or the proportionality principle may require discussions to be by telephone or video conference.

18.5 The parties, their lawyers and experts should co-operate to produce the agenda for any discussion between experts, although primary responsibility for preparation of the agenda should normally lie with the parties' solicitors.

18.6 The agenda should indicate what matters have been agreed and summarise concisely those which are in issue. It is often helpful for it to include questions to be answered by the experts. If agreement cannot be reached promptly or a party is unrepresented, the court may give directions for the drawing up of the agenda. The agenda should be circulated to experts and those instructing them to allow sufficient time for the experts to prepare for the discussion.

---

[5]   *Daniels* v *Walker* [2000] 1 WLR 1382.

18.7 Those instructing experts must not instruct experts to avoid reaching agreement (or to defer doing so) on any matter within the experts' competence. Experts are not permitted to accept such instructions.

18.8 The parties' lawyers may only be present at discussions between experts if all the parties agree or the court so orders. If lawyers do attend, they should not normally intervene except to answer questions put to them by the experts or to advise about the law.[6]

18.9 The content of discussions between experts should not be referred to at trial unless the parties agree (CPR 35.12(4)). It is good practice for any such agreement to be in writing.

18.10 At the conclusion of any discussion between experts, a statement should be prepared setting out:

(a) a list of issues that have been agreed, including, in each instance, the basis of agreement;
(b) a list of issues that have not been agreed, including, in each instance, the basis of disagreement;
(c) a list of any further issues that have arisen that were not included in the original agenda for discussion;
(d) a record of further action, if any, to be taken or recommended, including as appropriate the holding of further discussions between experts.

18.11 The statement should be agreed and signed by all the parties to the discussion as soon as may be practicable.

18.12 Agreements between experts during discussions do not bind the parties unless the parties expressly agree to be bound by the agreement (CPR 35.12(5)). However, in view of the overriding objective, parties should give careful consideration before refusing to be bound by such an agreement and be able to explain their refusal should it become relevant to the issue of costs.

19. Attendance of Experts at Court

19.1 Experts instructed in cases have an obligation to attend court if called upon to do so and accordingly should ensure that those

---

[6]   *Hubbard v Lambeth, Southwark and Lewisham HA* [2001] EWCA 1455.

instructing them are always aware of their dates to be avoided and take all reasonable steps to be available.

19.2 Those instructing experts should:

(a) ascertain the availability of experts before trial dates are fixed;

(b) keep experts updated with timetables (including the dates and times experts are to attend) and the location of the court;

(c) give consideration, where appropriate, to experts giving evidence via a video-link;

(d) inform experts immediately if trial dates are vacated.

19.3 Experts should normally attend court without the need for the service of witness summonses, but on occasion they may be served to require attendance (CPR 34). The use of witness summonses does not affect the contractual or other obligations of the parties to pay experts' fees.

© June 2005

Appendix 3

Example Conditions of Engagement

Expert Valuation Evidence

The Credibility Partnership
Independence Row
Barcester   5 January 2005

Dear Mr Andrews,

Imprudent Banking plc v *Fastbuck & Partners (a firm)*
Prime Location House — Valuation evidence

Conditions of engagement

Thank you for your letter of 2 January 2005 confirming your instructions to me for the provision of expert valuation advice and evidence in respect of the above-named legal action. I confirm, as discussed, my conditions of engagement for the acceptance of this appointment as set out below.

Case background

It is understood that your clients, Improvement Banking plc, have issued a claim in the Chancery Division of the High Court against the above-named defendants, Fastbuck & Partners, in respect of a loan security valuation dated 10 July 2002 upon which your clients sought to rely. They are alleging failure to undertake the advice competently as a result of which they have suffered loss and are seeking damages.

I further understand that the valuation report was in respect of an office investment property in Barcester known as Prime Court House, against the security of which your clients provided a loan facility.

Instructions

I am instructed to provide my expert valuation opinion in respect of the freehold interest in the Prime Court House office investment as at 10 July 2002, when it was subject to an occupational lease to Hedgebets.com plc for a term of 20 years at a rent of £350,000 per annum. I am asked to provide my opinion on the following three bases:

- On the assumption (which was the fact) of the hedgebets.com covenant as leased.
- On the assumption that it was let to a prime office tenant covenant.
- On the assumption that the property was unlet.

I understand that Hedgebets.com plc went into receivership in October 2002.

Expertise

I confirm that the provision of the valuation advice and opinions is within my professional experience and expertise.

Claim history

You have advised me that a letter of claim has been submitted to the defendant within the terms of the professional negligence pre-action protocol to which a response was received. The solicitors acting for Fastbuck & Partners have confirmed that indemnity insurers, Hardluck Assurance plc, have been informed. I further understand that a statement of claim has now been submitted and you are awaiting a reply from the other side.

Compliance

In accepting and undertaking these instructions, I confirm my understanding that, within the Civil Procedure Rules ("CPR 35"), it is my duty to help the court in matters within my expertise and this duty overrides any obligation to those instructing me or by whom I am paid. I am to provide an objective unbiased view. My report will be independent and uninfluenced by the pressures of litigation. I will comply with the requirements set out in CPR 35, Practice Direction 35 and the Civil Justice Protocol for the instructions of experts to give evidence in civil claims, in particular:

- The report will be addressed to the court.
- The report will contain the substance of instructions received.
- I will include details of my qualifications and experience.
- I will summarise any range of opinions, setting out the substance of all facts and the source of all information.
- The report will contain a summary of conclusions, stating any qualifications to my opinions.

I will include a statement that I understand my duties to the court and this has been complied with and will continue to be complied with.

The report will contain a statement of truth as required in Practice Direction 35 (para 2.4), in the following form:

> I confirm that insofar as the facts stated in my report are within my own knowledge I have made clear what they are and I believe them to be true and that the opinions I have expressed represent my true and complete professional opinion.

I will also comply fully with the RICS practice statement and guidance note entitled *Surveyors Acting as Expert Witnesses* (2nd edition). The *Practice Statement* is mandatory for all chartered surveyors providing expert evidence. A copy of the *Practice Statement* and guidance note can be supplied on request.

In regard to the valuations themselves, I confirm that they will be undertaken in accordance with the practice guidance and definitions contained in the current *RICS Valuation Appraisal and Guidance Notes* ("The Red Book").

The Red Book does provide for any possible departures by agreement and no doubt you will let me know if this is believed to be necessary in relation to any issue. A copy of the relevant definitions, eg in respect of "market value", can be provided on request.

Conflicts

I have already undertaken a conflict check and the results were sent to you with my letter of 23 December 2004. I confirm that having undertaken these enquiries, I did not identify any current or past relationships or connections with the property, which might reasonably threaten my independence or suggest a potential for bias in my role as expert witness for the claimant. You will have observed that the building consultancy department of my practice undertook a structural survey for Hardluck Assurance plc three years ago on a property in London. This is not perceived to give rise to any adverse implications in this context.

Confidentiality

I fully accept and acknowledge that the substance of this action, the identity of the parties and other relevant details and matters which should be kept entirely confidential will not be disclosed to any third party during the currency of my appointment, except in relation to any approved disclosure or in court proceedings.

Role and adviser

It is understood that your clients may require me to provide some assistance as adviser (rather than expert) in the early stages of the proceedings in regard to the pursuance of the claim. I confirm that I am able to provide this role so long as any potential conflicts or difficulties are acknowledged as and when these may arise and which may inhibit the degree to which the role can be undertaken. I confirm that this will not change my overriding duty to the court if, ultimately, a report is submitted. My role as adviser must take into account the possibility of conflict and no doubt you will inform me of any possible disclosure implications.

Timetable and court directions

It is understood that a case management conference has not yet been held at this stage of the proceedings and no formal timetable has been set down for this action. It is agreed that you will consult me in advance before any dates are agreed or are to be ordered so that I can check availability and respond accordingly. Based on my present state of knowledge of this case, I believe that my report in first draft could be made available by the end of February 2005.

Documents

You will be supplying me with all relevant case documents, including relevant correspondence, reports, any leases, title documents and drawings, together with copies of all "pleadings". It is further understood that you will be supplying me, in due course, with relevant witness statements of fact and any other relevant expert witness reports together with relevant documents disclosed by the other party.

Conferences and meetings

I will be available to attend any conferences with counsel in regard to the progress of the case and prior to the submission of my expert's report. You have mentioned the possibility of mediation and I accept that, as an adviser, I may be called upon to assist in any mediatory attempts to settle. I would recommend that a meeting with the other side's valuation expert, when appointed, is held early in the proceedings rather than later and certainly before an exchange of experts' reports since significant cost savings can often be made if the facts and issues are narrowed and agreed at an early stage in a case of this kind.

Inspections

Please could you supply me with necessary contacts to arrange an inspection of the subject property and any known limitations of the inspection that might be possible.

Valuation team

As already discussed, I will be employing a junior qualified valuer to assist me in the collation of information and data necessary for the preparation of my valuations. The valuer is John Haynes and he can be contacted on telephone extension 2520 within my office. I confirm, however, that the production of the valuations, the inspection of the selected comparable evidence, the preparation of my report and all other tasks as set out above will be undertaken by me.

Fees

As already discussed and agreed over the telephone, all of the work will be undertaken on a time basis at the following rates:

Martin C Farr   £200 per hour
                     (£220 per hour for court attendance)

John Haynes: £90 per hour

It is approximately estimated at this stage that the total cost of the report alone is likely to be £7000 (exclusive of expenses and VAT). This cost estimate is given for guidance only and will not be taken as a firm quotation. I am able to let you have a full estimate of all likely time costs that may be incurred if required.

The above fees are exclusive of all reasonable expenses incurred and VAT. There will be no calcellation charges.

I understand that you are likely to be billing your clients on a quarterly basis and have asked me to submit my invoices 10 days prior to each of those dates addressed to yourselves rather than the bank. Our invoices are payable 28 days from the date rendered.

The above time rates are open to annual review on the anniversary of this appointment.

We look forward to working with you, other professional advisers and your clients in regard to this appointment.

Yours faithfully,

John W Steadfast FRICS
Partner
The Credibility Partnership

Appendix 4

Expert's Report — Example Headings and Compliance

This is based on a fictitious negligence claim from Conditions of Engagement in appendix 3.

Title page

In the High Court of Justice   Claim No: HC02C03502
Chancery division

Between
Imprudent Banking plc (claimant)
and
Fastbuck & Partners (a firm) (defendant)

The expert evidence
of
John W Steadfast FRICS
of
The Credibility Partnership

provided to the Court

On instructions from
The claimant
Dated 5 November 2004

Contents and index

Report.
Appendices.
With paragraph and page numbers.
Repeat claim reference on each page.

Synopsis (executive summary)

Name, qualifications and experience (brief).
Background and context.
Instructions.
Brief description of property.
Other material facts (eg lease details).
Valuations.
Conclusions.

Experience and qualifications

Relevant to evidence, personal CV given in the appendix.
Commensurate with nature and complexity (Civil Justice Council Protocol 13.6.)
Give details of other opinions relied on with qualifications of originator (Civil Justice Council Protocol 13.8).

Instructions

Only need to include the "substance" of instructions received (CPR Part 35.10). But must not be incomplete or tend to mislead (Civil Justice Council Protocol 13.15).

Investigations, enquiries, documents seen

Statement of agreed facts in appendix (if relevant).
Details of documents, reports, literature and other material relied upon (PD 35 2.2(3)). Whether facts within own knowledge (PD 35 2.2(4)).

Background and history

Chronological summary of events and general background to the opinions sought (CPR Part 35).

The property

Situation.
Description.
Floor and site areas.
Lease synopsis (full heads of terms in appendix).
Rating assessment(s).
Service charges (if relevant).
Planning history and permissions.
Other relevant facts.
(Keep facts separate and discrete from opinion evidence.)

Market background and evidence

Discuss/begin valuation "trail".
Indices in appendix.

Comparable transactions and facts

Give details of those on which you rely (full list in appendix).
Indicate where not agreed with supporting data or documents.
Rents.
Investment yields.
(Clearly distinguish between those facts known to be true and those which are assumed.)

Valuations and advice

Explain method, basis, technical terms and inputs.
Give factors of influence.
Choice of rental and yield comparables.
Follow and complete valuation "trail".
Include synopsis of calculations.
Explain any assumptions and justify.
If necessary, include more detailed calculations in the appendix.

Potential for differing opinions

Fully explore the range of possible views and variability (Civil Justice Council Protocol 13.11). Where no source for range of opinion, it may be necessary to express a view as to the range of opinion which the expert might have arrived at if asked. But do emphasise the basis of the expert's judgment (Civil Justice Council Protocol 13.13).

Explain factors and results.

Final comments on realistic assumptions.

Give reasons why your opinions are to be preferred.

If material facts in dispute, express separate opinions on each hypothesis.

Respond to all instructions.

Conclusions

Final conclusions on issues as instructed.

Do not make any judgment comments (on the trial issues).

Declarations and compliance

My report has been prepared in accordance with:

- Part 35 of the Civil Procedure Rules, including the *Practice Direction* (PD 35).
- The Civil Justice Council Protocol.
- The *RICS Practice Statement and Guidance Note for Surveyors Acting as Expert Witnesses* (2nd edition).

In particular, I would confirm my understanding that:

- I have an overriding duty to help the court on the matters within my expertise and I confirm I have complied with that duty.
- This duty overrules any obligation to the person from whom I have received instructions or by whom I am paid.
- I confirm that insofar as the facts stated in my report are within my own knowledge I have made clear what they are and I believe them to be true and that the opinions I have expressed represent my true and complete professional opinion.

- My report includes all facts that a surveyor regards as being relevant to the opinion I have expressed and I have drawn attention to any matter that would affect the validity of that opinion.
- My report complies with the requirements of the Royal Institution of Chartered Surveyors as set down in the *Surveyors Acting as Expert Witness Practice Statement* and Guidance Note (2nd edition).

Signed:. . . . .

Dated: . . . .

Signed (witness): . . . .

Dated: . . . .

Appendix 5
Table of Cases

Appendix 6
Useful Addresses and Websites

RICS — Dispute Resolution
Surveyor Court
Westwood Way
Coventry CV4 8JE
Tel: 020 7222 7000
Website: *www.drs@rics.org*

RICS — Dispute Resolution Faculty
(includes Expert Witness Registration Scheme)
12 Great George Street
London SW1P 3AD
Tel: 020 7222 7000
Website: *www.dr.faculty@rics.org.uk*

Expert Witness Institute
Africa House
64–78 Kingsway
London WC2B 6BD
Tel: 0870 366 6367
Website: *www.ewi.org.uk*

Society of Expert Witnesses
PO Box 345
Newmarket CB8 7TU
Tel: 0845 702 3014
E-mail: *helpline@sew.org.uk*

UK Register of Expert Witnesses
J S Publications
PO Box 505, Newmarket
Suffolk CB8 7TF
Tel: 01638 561 590
Website: *www@jspubs.com*

Bond Solon
13 Britton Street
London EC1M 5SX
Tel: 020 7253 7053
Website: *www.bondsolon.com*

Thomson (Sweet & Maxwell)
Sweet & Maxwell Ltd
100 Avenue Road
Swiss Cottage
London NW3 3PF
Tel: 020 7449 1111
Website: *www.sweetandmaxwell.co.uk*

The Academy of Experts
3 Gray's Inn Square
London WC1R 5AH
Tel: 020 7430 0033
Website: *www.academy-experts.org*

The Lands Tribunal
Procession House
55 Ludgate Hill
London EC4M 7JW
Tel: 020 7029 9780
Website: *www.landstribunal.gov.uk*

The Valuation Tribunal Service
Upper Ground Floor, Block One
Angel Square
1 Torrens Street
London EC1V 1NY
Tel: 020 7841 8700
Website: *www.valuation.tribunals.gov.uk*

The Civil Justice Council
Room E214
Royal Courts of Justice
London WC2A 2LL
Tel: 020 7947 6670
Website: *www.civiljusticecouncil.gov.uk*

Index